# The Challenge of Recycling Refrigerants

Kenneth W. Manz

**BNP** Business News Publishing Company
Troy, Michigan

**Copyright © 1995**
**Business News Publishing Company**

All rights reserved. Except as permitted under the United States Copyright Act of 1976, no part of this publication may be reproduced or distributed in any form or means, or stored in a database or retrieval system, without the prior written permission of the publisher, Business News Publishing Company.

---

**Library of Congress Cataloging in Publication Data**

Manz, Kenneth W., 1951-
    The challenge of recycling refrigerants / Kenneth W. Manz
        p. cm.
    ISBN 1-885863-00-4
    1. Refrigerants--Recycling    I. Title.
    TP492.7.M343    1995        94-34530
    621.5'64--dc20            CIP

---

Editors: Joanna Turpin, B.A. Checket-Hanks
Art Director: Mark Leibold
Copy Editor: Carolyn Thompson

This book was written as a general guide. The author and publisher have neither liability nor can they be responsible to any person or entity for any misunderstanding, misuse, or misapplication that would cause loss or damage of any kind, including loss of rights, material, or personal injury, or alleged to be caused directly or indirectly by the information contained in this book.

Printed in the United States of America
7 6 5 4 3 2 1

# Acknowledgments

When my work began on refrigerant recycling, I did not intend to write a book about it. Early and continuing influences in my life contributed to my preparation for this work.

First of all, I thank God who gave me opportunities, abilities, and continuing strength. I thank my father Elias and my mother Evelyn, whose strength of character on one hand and compassion on the other have shaped my life.

I thank my wife Gwen for her understanding and love in encouraging me and for keeping the home fires burning when I was gone so often. She has been my partner in intellect. I thank my children for keeping me young at heart even as my hair has grayed and thinned in the last nine years.

I thank my teachers at Paulding High School, particularly my math teacher Mr. Forrest Albert. I thank the professors at Purdue University, especially Mr. Johnson at the Fort Wayne Campus, and Professors Alan McDonald (fluid mechanics) and Frank Incropera (heat and mass transfer) at the main campus.

Three mentors in my professional career prior to refrigerant recycling stand out in my mind. Mr. Ed Wene, Manager of the Experimental Stress Analysis and Structural Dynamics Laboratory at International Harvester, challenged me and taught me technical writing. Mr. Ralph Lower, Vice President of Engineering at Robinair, hired me and taught me much from his vast experience in refrigeration system design. Also, Mr. Howard Pratt, Applied Mechanics Laboratory Supervisor at International Harvester, provided understanding and guidance to me when I thought the future couldn't come fast enough.

Time and space prevent me from enumerating all the colleagues and valued friends I have made in the last 10 years. Some of these people include Ward Atkinson, the veteran SAE Committee Chairman, Herb Gilkey, my ASHRAE mentor, and Sue Correll, who encouraged me to get my feet wet in ASHRAE contaminant control activities. My sincere thanks go to Mr. William Clogg, President of Robinair, for his support and encouragement to write this book. I also thank all my fellow management team members at Robinair, as we have shared in the wild ride to manage the unpredictable refrigerant service tool business.

This book would have been impossible without the following contributors: Mr. Richard Cawley of The Trane Company was invaluable in his criticisms, which were sincere and very helpful. Dick's understanding of me and encouragement to write this book were an inspiration. Mr. Herb Phillips, Vice President, Engineering, ARI, has worked at my side throughout the development of ARI 740 and ISO 11650R Standards. As a book reviewer, Herb's critiques were brief, to the point, and right on the mark. Ms. Deborah Ottinger of the U.S. EPA has clearly demonstrated that governmental regulators can work with industry in a cooperative way. Debbie has always listened, even when we often differed, and I consider her my colleague and friend. Her critiques provided a different perspective. Mr. Tom Mahoney of the *Air-Conditioning, Heating, and Refrigeration News* was brief and prompt in his remarks. Tom has battled for wider use of recycling and encouraged me to write the book. Mr. Shelvin Rosen reviewed the chapters on contaminants, and his contributions were helpful. My final external reviewer was Mr. Fritz Peacock of the Purdue University Herrick Laboratories. Fritz's experience in teaching at the vocational school level led to the inclusion of Chapter 11 and the summaries at the end of each chapter.

Nan Mitchell typed my manuscript, proofread every page, and assembled all the photographs. Thanks Nan for all you do in so many ways. My son Tony prepared all the computer-generated illustrations that make understanding the material so much easier.

My internal reviewers at Robinair were Jim Biedenharn, George Gardner, Sandy Sheehe, Ilya Reyzin, and Christopher Powers. Valuable design contributions came from Gary Murray, Roger Shirley, Gregg Laukhuf, Charlie Dull, and Dan Olds.

# About the Author

Kenneth W. Manz graduated from Paulding (OH) High School in 1969 and from Purdue University with a BS in Mechanical Engineering in 1973. He worked for International Harvester Motor Truck Engineering as an experimental stress analyst before joining Robinair Division, SPX Corp., in 1982. His current position is Director, Advanced Technology.

Mr. Manz has been active in refrigerant recycling since 1986. He has served as an active member of the SAE Interior Climate Control Committee and helped develop all of the SAE documents pertaining to refrigerant recycling and retrofitting that are referenced throughout the book. He has served on two ASHRAE committees since 1986 — TC 3.3, Technical Committee on "Contaminant Control in Refrigerating Systems and in Used Refrigerants" (Chairman 1992 to 1994) and TG2.HE, Task Group on Halocarbon Emissions (Chairman 1991 to 1993). As Research Chairman of TC 3.3 from 1988 to 1992, Mr. Manz led the efforts on ASHRAE research projects 601RP and 683RP concerning recycling and used-refrigerant analysis. He also was lead author in writing a new "Recovery, Recycling, and Reclamation" section and revising the "System Cleanup Procedure After Hermetic Motor Burnout" section in Chapter 6 of the *1994 ASHRAE Refrigeration Handbook*.

Since 1990, Mr. Manz has chaired the Air-Conditioning and Refrigeration Institute (ARI) Engineering Subcommittee, which writes and revises Standard 740, *Performance of Refrigerant Recovery and/or Recycling Equipment*. Since 1991, he has served as convener for the ISO working group that writes and revises Standard 11650R, *Performance of Refrigerant and/or Recycling Equipment*. He was appointed by ARI to serve on the Industry AD Hoc Group that wrote *Handling and Reuse of Refrigerants in the United States* in 1994. He also serves on the board of directors of the Alliance for Responsible Atmospheric Policy (formerly the Alliance for Responsible CFC Policy).

Mr. Manz has led the effort to relax recycling restrictions to permit contractors flexibility in handling refrigerants. He used the slogan, "Making a crime out of selling recycled refrigerant," as part of the National Recycling and Emissions Reduction Act (U.S. Clean Air Act Amendments Section 608) to help bring about the Sunset Clause, which limited the time period for which the sale of used refrigerant was prohibited, unless it had been reclaimed, to two years. He received "The

Stratospheric Ozone Protection Award for Individual Leadership in Refrigerant Recovery and Recycling" from the U.S. EPA in 1993.

Mr. Manz is a practicing engineer and inventor who currently holds 26 U.S. patents and many foreign patents in the fields of recovering, recycling, and recharging refrigerants. He has authored numerous technical articles on methods of recycling.

Mr. Manz and his wife Gwen live in rural Paulding, Ohio, with their 10 children.

# Preface

This book is targeted for contractors, service technicians, in-plant maintenance personnel who work on air conditioning and refrigerating systems throughout the world, and students who are learning to become the service personnel of the future. It will also be useful for design engineers, building owners, and the general public who either work in the air conditioning and refrigeration industry or wish to understand the economic impact of the phaseout of chlorofluorocarbon refrigerants.

This book raises issues relating to recycling refrigerants and makes comparisons to related recycling issues with other products. It discusses environmental issues, economics, government and industry responses, and how these responses affect our lives. No effort is made to mask the complexity of the issues. The emphasis is to paint the "big picture" as a consensus of the way industry sees it, so service personnel and students can take better actions and make improved service choices.

The book is not prescriptive — it does not instruct you to take certain actions and "ask no questions." Rather, it characterizes the controversy, the options, and brings together certain information that can help you choose the appropriate option. It covers the effects of inactivity and the benefits of some actions over others, in as realistic a manner as possible.

In this book, refrigeration industry participants each can be considered to have a hand in equipping and training service personnel with tools and information. Service personnel are the only ones who can make the decisions to purchase and use equipment and to employ the recommended service practices. This book attempts to tell the whole story and expose you to as many options as possible, so that, coupled with your own knowledge and experience, you can implement refrigerant conservation and recycling practices in ways we "engineers" could not even imagine. Hats are off to you, the foot soldiers of our industry, as you go out to implement society's demands for a safer environment.

The goal of this book is to prepare service personnel for major changes in refrigeration service, which have already begun to take place and will accelerate in the next several years. By this, I mean the complete replacement of the comfortable and familiar CFC and HCFC refrigerants on one hand and the widespread use of recycling and conservation practices on the other. The knowledge and skill required to properly service refrigerating systems will demand continuous education and improvement to keep up with new refrigerants, conservation practices, retrofitting practices, new technology, and the related issues of indoor air quality and energy efficiency. There are truly exciting and challenging years ahead.

The book focuses on cleaning refrigerants for reuse, keeping refrigerating systems clean, protecting the used-refrigerant supply, and analyzing how recovery-recycling equipment works to accomplish a variety of functions. While others have written about recovery practices and minimizing the time required, they have not concentrated on the equipment design and selection criteria to the extent this book does. While textbooks on air conditioning and refrigeration systems typically include an abbreviated chapter on service equipment and practices, this book offers a much more complete picture and a guiding hand to understand the rapid changes in acceptable service practices.

Historical events have characterized the development of recycling equipment and practices. A sense of what happened in development laboratories, committee rooms, and service locations is hereby preserved. The saga will continue to unfold as certain issues are resolved and problems solved; they will give way to other new issues. It is hoped that the contents of this book will result in discovery of even better design and service practice options.

Kenneth W. Manz

# Contents

Foreword .................................................................................................... i

## Chapter One
Recycling and Renewable Resources ..................................................... 1

## Chapter Two
Refrigerants — Now a Renewable Resource ......................................... 9

## Chapter Three
Refrigerant Recovery Equipment ........................................................ 17

## Chapter Four
Oils, Acids, and Particulates ............................................................... 45

## Chapter Five
Moisture ............................................................................................. 55

## Chapter Six
Air and Non-Condensables ................................................................ 73

## Chapter Seven
Mixed Refrigerants ............................................................................. 89

## Chapter Eight
Deciding How to Process Refrigerant for Reuse ............................... 105

## Chapter Nine
Choosing a Recovery-Recycling Unit for Stationary Applications ........ 115

## Chapter Ten
Motor Vehicle System Recycling and Retrofitting ............................. 137

**Chapter Eleven**
Service Procedures ............................................................................................... 161

**Chapter Twelve**
A Look to the Future .......................................................................................... 179

**Index** ................................................................................................................ 183

# Foreword

Mr. Kenneth Manz, author of *The Challenge of Recycling Refrigerants*, and the subject of refrigerant recycling have become synonymous terms. When one is mentioned, the other is brought to mind. Indeed, Mr. Manz was submerged into recycling technology many years before ozone-depletion concerns and subsequent regulations were accepted facts. His expertise was eagerly solicited by engineering societies, trade associations, the U.S. Environmental Protection Agency, and others as research was conducted and regulations and standards were being prepared. He gave of himself tirelessly to these organizations.

He also assisted (and continues to assist) many individual companies and service groups as they have prepared to meet the challenge of new standards and regulations for preserving refrigerant and protecting our environment. The air conditioning and refrigeration industry — and society as a whole — is indebted to Ken Manz.

This book is timely. It is particularly refreshing because it is not prescriptive in nature. Options exist in spite of heavy standards and regulations, and Mr. Manz carefully explains these options so that the reader can choose recycling equipment and service tools and procedures that best accomplish his objectives, while remaining responsible within the framework of laws and standards as a result of environmental concerns. Mr. Manz also commends and issues challenges to service technicians for their important roles in preserving the quality and integrity of current and future refrigerants.

Other books will undoubtedly be prepared with material similar to that in *The Challenge of Recycling Refrigerants*. While such books may be of interest, please bear in mind that Mr. Manz has been dedicated and involved with refrigerant recycling and preservation for many years. As mentioned earlier, his involvement precedes the initial "ozone hole" scare of 1987. He is regarded by his colleagues as a tenacious worker of highest character. His many patents and research projects, and his involvement with product development and the preparation of industry standards, are manifestations of his drive. He is one who makes decisions by asking, "Is it the right thing to do?"

Richard E. Cawley, P.E.
Manager, Systems Technology Engineering
Unitary Products Group
The Trane Company



The theory that the life-protecting stratospheric ozone layer is being destroyed by chlorinated refrigerants released to the atmosphere has created the greatest upheaval ever imposed upon the refrigeration and air conditioning industry. That upheaval has been the greatest challenge the industry has had to face in its history.

As the industry, its technology and, thus, its products have advanced, the demands and needs for the products have become critical. User dependence on this technology has evolved to the point where the products in a number of applications (e.g., the fields of medicine and food preservation) have become a necessity rather than a luxury. Therefore, the challenge has become so much greater, and the need for resolution so much more critical, than had the theory been exposed earlier in the industry's history.

The challenge has created new industries and has expanded others. Refrigerant reclamation, the manufacture of recovery and recycle equipment and refrigerant-identifying equipment are examples. More pressure gauges, hoses, fittings, and recovery and storage tanks are being manufactured and sold than ever before. New refrigerants must be developed and manufactured.

A major and critical link in addressing this upheaval and resolving the challenge is the proper handling of refrigerants. Whether it be simple recovery, transfer from one container to another, recycling, etc., an understanding is essential of how all of this must be done, why, and the equipment available for its accomplishment.

Ken Manz has carefully described these important elements in the first book of its kind. It is "must" reading for anyone and everyone involved.


Herbert Phillips, P.E.
Vice President, Engineering
Air-Conditioning and Refrigeration Institute

# Chapter One

# Recycling and Renewable Resources

Recycling and other measures that protect the environment raise complex issues. Individual rights often conflict with responsibilities to others. Societal groups hold diverse opinions as to which problems are serious, who should bear the associated costs, and the benefits of recycling. Should compliance be voluntary, encouraged economically, or enforced? In this chapter, these and other issues are discussed in order to understand the complexity behind individual and societal decisions to recycle.

## The Recycling Mindset

People take care of things they value and cannot easily replace. The value placed on an item depends on:

- interests and needs of the individual user;
- cost and availability of a new product or item;
- demand for and extent of interest shown by other users;
- resale value or cost of disposal;
- the individual's perception of how others will view owning and caring for the item;
- regulations governing the product's use.

Societies have long placed value on precious metals, historical artifacts, and items of rare beauty. They place value on medicine and procedures to heal the human body. People desire products and services that will improve convenience and their lifestyle. Examples are the automobile, plow, air conditioner, and dishwasher.

Most people find it difficult to deal with the hidden effects of their choices. In many countries, people tend to think about the new product more than the disposed product. They mentally visualize a product's use while new, more than when it is worn out. Many do not think of who might use a product after them or the effect downwind or downstream.

Environmental issues deal with such hidden effects. Users often become polluters through improper use or disposal, yet they are not aware of the consequences of their actions on others. Both time and place may separate polluters from the effects of their pollution.

We are all placed under tremendous pressure to address environmental issues. Societies, not merely governments, should take the following actions:

- Conduct research to determine the causes of environmental damage, reasonable objectives, and benchmark measurements whereby improvements can be monitored.
- Encourage technological solutions for new products and processes to address pollution, recycling, and disposal, while encouraging the best minds to work on the most serious problems.
- Encourage all affected people to contribute their individual efforts.
- Take steps to respond to environmental concerns before fully conclusive evidence is established, if available evidence points to serious problems.
- Encourage cooperation between producers, users, the public, and governments to solve problems, minimize disruptions and costs, and protect the environment.

Societies must avoid the temptation to:

- view every environmental issue as a crisis;
- act without evaluating the effect of the replacement product or process;
- emphasize government actions above cooperative and individual efforts (people acting responsibly will minimize the temptation and need for this).

People's interests and values are associated with their beliefs, knowledge, and experiences. These values vary considerably both between individuals and between societies. Natural events and new discoveries often change people's perspectives.

In response, some resist change while others impulsively embrace it. Those who continuously cry "wolf" will not be heeded. Those who resist all change will see the iceberg of their changeless world melt away. The task remains for diverse groups to work together, to investigate, and appropriately respond to change.

A recycling mindset exists when an individual or society has investigated a product and become convinced of the value of taking care of it or using it wisely. They have considered it, not only from their own perspective, but also from others. They are prepared to act responsibly as good stewards. After the recycling mindset is established, the "how" of recycling is a smaller challenge.

As used in Chapters 1 and 2, the term *recycling* is the care of products while operating, and the collection and reconditioning of products or materials for reuse so as to minimize the environmental effect when disposing of used products.

# Consumable Product Life Cycle

Consumable products are used up after a limited time and then need to be replaced. Examples include the following:

- Paper plates, cups, and napkins
- Disposable surgical gloves and syringes
- Engine oils and other vehicle fluids
- Inexpensive calculators, watches, and even cameras
- Paper bags and cardboard boxes
- Batteries

Even land, usually considered a renewable resource, has been treated as consumable. In the 18th and 19th centuries, southern U.S. cotton growers planted the same crop year after year, until the nutrients and soil condition were depleted. Then they abandoned the land to move to land that had not previously been farmed. Today, this pattern is repeated in the rain forests.

The consumable product life cycle consists of manufacturing, using, and discarding. This cycle is often associated with a "disposable mindset" or "throw-away society." Manufacturers, concentrating on the point of purchase, package the product attractively and feature the use of the product while it is still new.

The user is interested in price, convenience, and fulfilling his needs again with emphasis on the new product. When the user is finished with the product, he either abandons it, throws it in the trash, or sells it to someone else so it can be their problem.

Disposing of consumable products is a major problem. If everyone used paper plates and threw them in the trash, the landfills could not keep up. Too often, neither the price of the product nor the purchase decision take the costs of disposal into account at all.

Beverage cans are another example. They are often found along the roadside, creating an eyesore and making it difficult to mow the grass. In the environment, they can provide undesirable places for breeding disease-carrying mosquitoes. Organizations, such as the American Farm Bureau, lobbied for a can deposit in some states so offenders would pay the deposit and enterprising citizens would collect the cans and receive the deposit. If more citizens would act responsibly, deposits would not be required. Meanwhile, the deposits help produce the desired effects of keeping the roadsides clean and recycling the cans.

The pollution generated while using a product is not considered by those having a disposable mindset. They only consider important direct effects on them, their loved ones, or their property. The hidden damage to others is "out of sight, out of mind."

There is a valid need for some consumable products. Without gasoline or electricity, cars and factories could not run. Without food, the human body would not function. There is, however, no room for a disposable mindset.

People who have a disposable mindset are not "wicked;" they are either unaware or thoughtless. When presented with evidence of the harm they have caused, or when there are laws to protect society, those who ignore them are criminals. The motto, "Do unto others as you would have them do unto you," also applies to their land, water, and air.

## **Renewable Product Life Cycle**

Renewable products are those that are used for a relatively long period of time and then recycled for reuse. Examples include the following:

- Glass plates and cups
- Draft horses and other beasts of burden
- Objects made from precious metals
- Water
- Forests and trees

The Native Americans used the American Bison (buffalo) in a renewable way. They did not own the buffalo, as people today would define ownership; individuals could not point out just which buffalo in the herd was theirs. But they used the hide for tents, the meat for food, the tallow for light, and the dung for fuel. They took what they needed, and the buffalo herd was maintained.

In contrast, the buffalo was entirely expendable to the pioneers who settled the western United States. They brought their cattle and their seeds with them, often killing the buffalo for sport and leaving them on the prairie, or taking only the choicest cuts of meat. The pioneers exhibited many admirable qualities, but they were inconsiderate of the Native Americans and their ways. It is doubtful whether the Native American lifestyle could, by any means, have been preserved in the face of progressive civilization. Yet, Native American views on renewable resources and recycling were advanced. Much can be learned from studying them.

Organic farming is another example of managing renewable resources. Dried plants, unused food, and other organic refuse are mixed and stirred with soil to create nutrient-rich and humus compost. This can be spread over a garden to improve fertility. Animal manure can be spread over the land and mixed with the soil for the same purpose. Spreading grass clippings between rows of garden vegetables helps retain moisture and keeps down weeds. Organic farming increases the activity of earthworms, which help convert soil nutrients to a usable form and improve drainage.

The renewable product life cycle consists of manufacturing, using, maintaining, recycling, and safe disposal. It is associated with a recycling mindset, which considers the entire life cycle. Manufacturers adopting this philosophy concentrate on efficiently using raw materials, minimizing operating energy requirements, and maintaining their product. They consider its effect on the environment and user safety and include methods for recycling and disposal. Of course, they must provide products that work in the application for a competitive price.

Users with a recycling mindset purchase a product that will last and consider usage over the entire planned period, the resale value based on condition, and disposal of the final product. They practice recycling and pollution control. Even in the absence of regulations, the manufacturer and user possessing the recycling mindset do all they can, and know how to do it.

People often base the value of an item not only on their need, but on the cost to replace it and how far they have to drive or walk to get the replacement. This often spawns carelessness among affluent individuals and societies. For example, people living in warm climates probably would not miss a pair of gloves. Others living in cold climates can usually replace lost gloves for less than one hour's pay. Still others in cold climates would pay a week's wages if there were any gloves to be found. For many, the economics of price and supply dictate behavior, rather than caring for their gloves.

Societies should not measure an individual's success by the number of garbage trucks that pull away from his house each week. People work to improve their standard of living. The more they possess, the greater responsibility they have to use these goods in a way that also improves the living conditions of others. Therefore, responsible people should try to minimize the effect of their way of life in consuming natural resources and on the environment.

## **Product Values Change**

During the 1970s, gasoline prices rose dramatically. For years, Americans drove large cars with poor fuel economy. When gasoline prices rose, people bought smaller cars, used public transportation, car pooled, and stayed home more. Whether in response to long lines and longer hours at the gas pump, price, or dependency on foreign oil, people viewed this commodity with new appreciation.

Similar response occurs when scrap metal prices rise. People begin to turn in their rusty nails or aluminum cans. This reduces the amount of trash going to the landfill or left along the roadside. Recycling often leads to lower production costs for aluminum and steel. Even when the price is higher, the total societal cost may be reduced.

Water is a precious resource in many regions. Whether from lack of rainfall or the cost of purifying the available supply, some regions are sparsely populated due to poor water supply. Droughts in populated areas, such as southern California, have heightened awareness and changed consumption habits. In most parts of the world, bottled water of known purity commands a premium over the normal drinking water.

## Used Materials

Consumers desire uniform product quality. This is one of the secrets of McDonald's hamburger success. A "Big Mac" in Paulding, Ohio, tastes the same as one in Moscow, Russia.

Uniform quality in used parts is sometimes critical. People often prefer factory reconditioned or rebuilt automotive parts over used parts of unknown quality. Experienced mechanics may be able to inspect or measure used parts to select the right ones, but the inexperienced motorist is well advised to solicit the advice of a mechanic or select new parts.

The market for used parts depends on:

- cost versus new;
- ease of recovery (used fenders versus used pistons);
- availability of new parts;
- value as part versus value as scrap metal.

At auctions everywhere, people buy anything from used furniture to used bicycles to used lawnmowers. At garage sales or flea markets, they buy used shoes, toys, and dishes. By private sale, anything one person is willing to buy and another is willing to sell at a price they can agree on, can and will be sold. The terms of sale for these items is usually "as is," meaning there is no guarantee.

Some items offer a variety of terms of sale. For instance, coins may be offered ungraded, graded by the dealer, or graded by a third-party grading service. A better guarantee usually means a higher price.

In short, it is well established that used parts may be sold in different conditions at various prices. Parts kept in better condition and those offered with warranty or grading usually bring higher prices. It often takes some time after new products are introduced to establish a used market. Reputable dealers soon become known. Natural market forces handle most products.

When fluids, such as cleaning solvents or engine coolants, are recycled and resold, the general public may not be able to distinguish the contaminated from the uncontaminated. Cleaning and testing standards may be set by manufacturers' associations or professional societies. These voluntary standards are not aimed at restricting sales, but rather serve to describe product quality so a fair price can be determined. (The example of recycled refrigerants in motor vehicles is discussed in Chapter 10.)

If an automotive mechanic knew that engine oil would be recycled for reuse, he would be cautious about adding oil-treatment additives. While the additive may be beneficial in the one car, it would become a contaminant to the general used oil supply. Another example is soda pop. Different people may like orange, cola, or raspberry; it would be difficult to sell a mixture of the three.

This becomes an important axiom when dealing with recyclable fluids. Before adding anything to a particular fluid or system, consider the effect of the additive on that system, the recycled fluid supply, and ultimate disposal of the fluid. Often, recycled fluids are handled by service technicians or other skilled personnel. When the recycled fluid is first introduced, the technicians may require special training for handling them. The sale of used fluids becomes less of a concern when handled and introduced into systems by trained, skilled technicians.

## Summary

Recycling raises complex issues. Studying these issues helps to deal with them responsibly. The recycling mindset implies taking care of products and using them in a way that benefits others. The disposable mindset concentrates on short-term needs and discarding used products for others to take care of. Renewable products and life cycles emphasize care and recycling. Disposable products and life cycles emphasize convenience and use while new. Environmental impacts are often hidden, and cause-effect relationships are indirect.

People often use items carelessly that are easily replaced. Their habits and perceptions change when the supply becomes limited. While initial acceptance may be limited, markets for used products develop. Recycled fluids may require voluntary standards to describe their quality. Before adding anything to recyclable fluids, the effect on the recycled fluid supply must also be considered. Special care may be required to preserve the quality of the recycled fluid.

## Chapter Two

# Refrigerants — Now a Renewable Resource

Refrigeration systems have always been treated as renewable, in that repairs are normally made and they have a long life cycle (up to 30 years). However, refrigerant has been treated as consumable. Whenever a pressurized system was opened for service, the refrigerant was almost always vented to the atmosphere.

When Thomas Midgely introduced the chlorofluorocarbon (CFC) refrigerants in the early 1930s, they were billed as wonder chemicals. They were colorless, odorless, non-toxic, and non-flammable. CFCs made excellent refrigerants, solvents, blowing agents, and aerosol propellants. As refrigerants, CFCs overcame the obvious shortcomings of $SO_2$ and ammonia, both of which have offensive odors and under certain situations can be lethal. CFCs were inexpensive to produce and in abundant supply. Finally, CFC refrigerants were chemically stable, lasting indefinitely under the expected operating conditions.

Ironically, the three wonderful characteristics of CFCs proved their undoing. Their apparent *non-toxicity* led to disregard for careful handling and containment of CFCs. This led to venting practices that would not have occurred if CFCs were associated with human health or environmental concerns. The only known health concern was possible asphyxiation caused by large leaks in unventilated areas.

The *low cost* of CFCs encouraged venting and sloppy leak repair practices. If the price of new refrigerants had been higher, recycling would have been practiced more commonly.

The *long life* of CFCs permitted migration to the stratosphere, where special environmental conditions created ozone depletion. Had they been less stable, CFCs would have broken down in the troposphere and not have disturbed the ozone layer.

When servicing automotive air conditioners, mechanics and "do-it-yourselfers" commonly added one or more 14-oz (0.4-kg) cans of refrigerant each year, rather than fixing the leaks. When repairing refrigerators and small air conditioners, technicians used a sweep-charge method to remove air and moisture, which resulted in venting even more refrigerant. CFC refrigerant used for leak testing also was vented.

With the exception of low-pressure refrigerants, such as R-11, and very large systems containing 500 lb (230 kg) or more refrigerant, refrigerant has always been vented.

Because most refrigerant was vented during service, new refrigerant was charged into most new or repaired refrigeration systems. For larger systems, the refrigerant was isolated in one portion of the system while repairing another portion. Some systems had isolation valves or pumpdown units and separate tanks for this purpose. R-11 was pumped into open containers during service, then pumped back into the same system or another system on the same site.

The building owner and service contractor were primarily concerned about cleaning up the contaminants left in the refrigeration system, whether they vented, isolated, or removed the refrigerant. They were not generally concerned with the quality of new refrigerant, or with moving refrigerant from system to system. Yet, some contractors practiced limited recycling. Other contractors introduced a small amount of a second refrigerant into a system to enhance performance or improve oil return. Since the refrigerant was vented, rather than introduced into another system, the only concern was the effect on the performance and durability of that system.

## Disposing of Components and Fluids

In the past, used compressor oil was poured on the ground, burned, or sent to the landfill. Used filters and other components were normally discarded in the trash. Contractors used to be guided by the most economical and convenient way to dispose of used product. More responsible contractors now dispose of used oil in a more environmentally responsible way.

A market for used systems has developed. Chillers and other large pieces of equipment are dismantled and moved; compressors and other components are sometimes remanufactured; and refrigerators, freezers, and window air conditioners are sold to other consumers. Automotive air conditioners are less frequently sold intact, due to their high cost of installation. Automotive salvaging yards have taken the lead in "parting out" or scrapping vehicles and automotive air conditioners.

## Ozone Depletion and the Role of Refrigerants

Atmospheric scientists have discovered a thinning of the earth's protective ozone layer. This layer is thin and dispersed; it can better be associated with a wisp of smoke than a blanket. It is located in the stratosphere, which is located 33,000 to 160,000 ft (10 to 50 km) above the earth. (Changing atmospheric weather patterns shift the ozone layer, so it is difficult to model or even measure at any one altitude.) The ozone layer serves to block ultraviolet (UV-B) radiation from the sun. Higher levels of UV-B radiation cause more cases of skin cancer and cataracts.

Drs. Sherwood Rowland and Mario Molina were prominent among discoverers of ozone depletion. Since that time, scientists have conducted research which many believe has associated ozone depletion with CFCs.

In a nutshell, scientists argue that the rise of CFCs has matched the rise in chlorine concentration in the stratosphere, which in turn has been associated with a decrease in ozone. This has not been without serious debate, in which some prominent scientists have argued that ozone depletion is not caused by CFCs. Scientists have measured the total column ozone amounts from ground-based, satellite-based, and aircraft-based instrumentation. They have confirmed that reductions in stratospheric ozone levels have occurred and continue to occur over most of the earth.

The ozone depletion debate may be represented in the following timeline:

- 1974 — Rowland & Molina first develop theory that CFCs could destroy stratospheric ozone.
- 1975 to 1986 — Atmospheric models predict that slight ozone depletion will occur by 2050. None yet observed.
- 1986 to present — Ozone depletion observed repeatedly. Predictions and models revised to account for accelerated depletion.

In the stratosphere, CFCs and HCFCs destroy ozone through chemical reaction. Ozone ($O_3$) is a chemical compound consisting of three oxygen atoms in each molecule. It is a relatively unstable gas compared to the oxygen molecule ($O_2$), which contains two atoms, Figure 2-1.

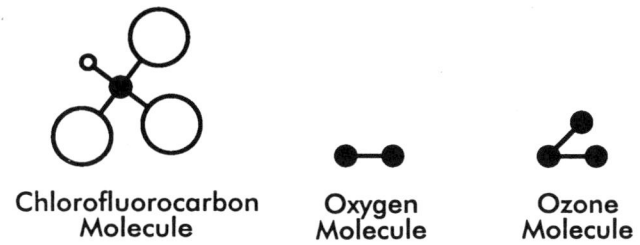

*Figure 2-1. CFC, $O_2$, and $O_3$ molecules*

The intense sunlight in the stratosphere causes a chlorine atom to break away from a CFC or HCFC molecule, Figure 2-2. This free chlorine atom joins with one of the oxygen atoms in an ozone molecule to make chlorine monoxide and a diatomic oxygen molecule. As free oxygen atoms are formed, they join with the oxygen in the chlorine monoxide, releasing the chlorine to repeat the chain reaction with other ozone molecules.

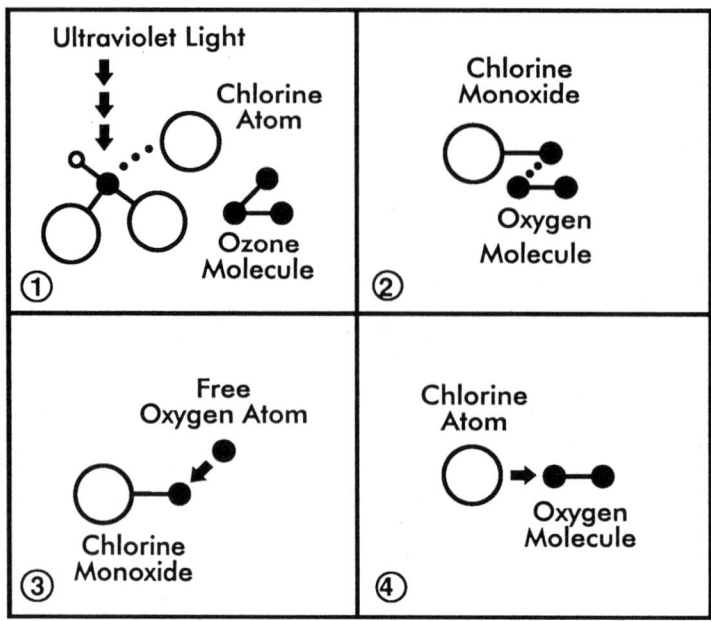

*Figure 2-2. Chemistry of chlorine atoms in the stratosphere*

## Montreal Protocol

In 1987, an historic treaty was signed by 24 countries under the United Nations Environmental Program. This treaty was called the Montreal Protocol on Substances that Deplete the Stratospheric Ozone Layer. The original protocol called for a 50% reduction in CFC chemical production by 1998, Figure 2-3.

The Montreal Protocol was the first cooperative effort by nations from around the world to protect the environment and call for reduction and eventual elimination of a class of compounds. It called for periodic scientific, technical, and economic options assessments to determine if further revisions were required.

The Montreal Protocol has been revised twice and now calls for elimination of CFC refrigerants by 1996 and for the reduction and eventual elimination of HCFC refrigerants by 2030. The third assessment will be completed in late 1994; further acceleration of HCFC phaseouts can be expected.

## 1990 U.S. Clean Air Act Amendments (CAAA)

Congress passed the 1990 CAAA in November, 1990. These called for the phaseout of CFC and HCFC refrigerants to a schedule no less stringent than the latest revision of the Montreal Protocol. The U.S. Environmental Protection Agency was given responsibility to draft and enforce regulations implementing the 1990 CAAA.

Refrigerants — Now a Renewable Resource

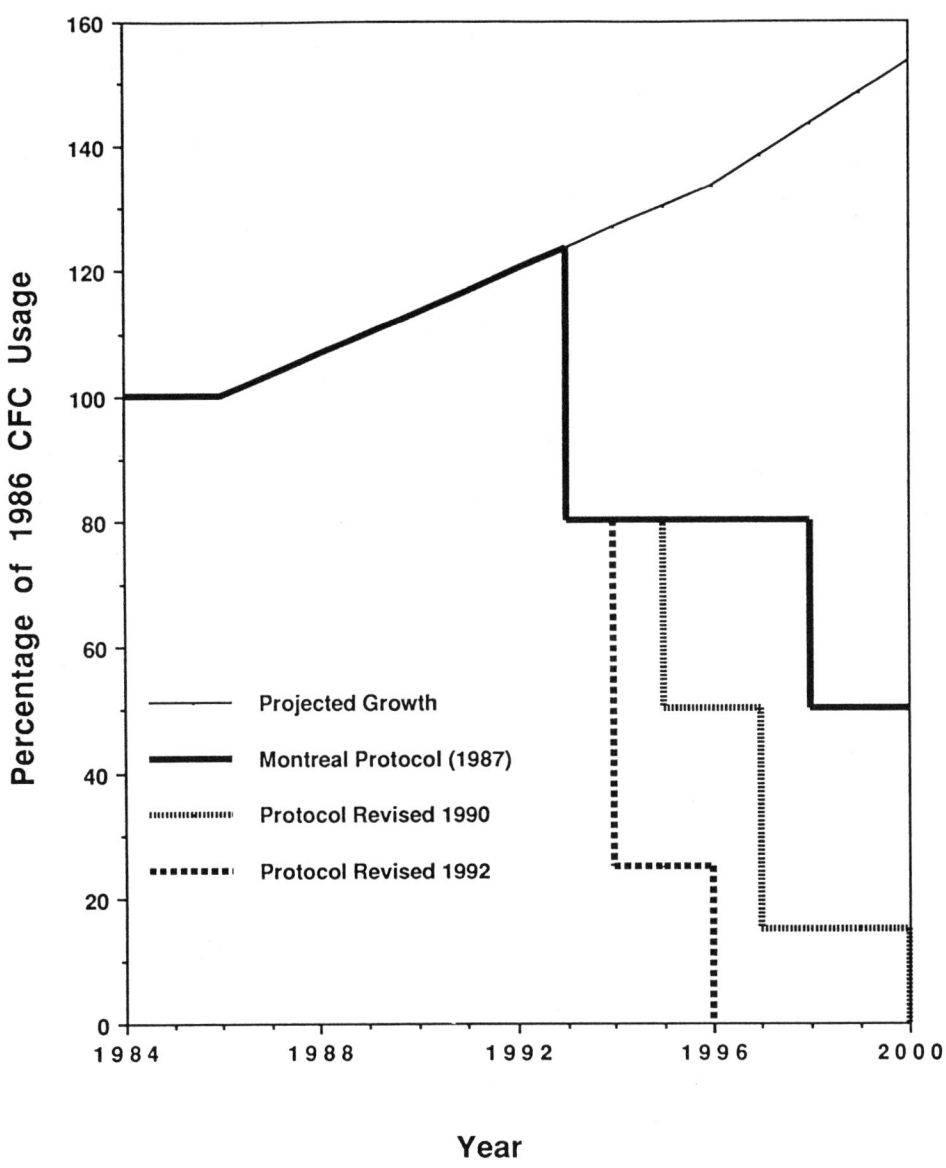

Figure 2-3. CFC phaseout

***Section 608 regulations*** — This section is called the *National Recycling and Emissions Reduction Program*. It applies to persons installing and servicing all air conditioners and refrigerating equipment except motor vehicle air conditioners (MVACs). Section 608 covers disposal of all equipment, including MVACs. The National Recycling Program has four basic provisions:

- Reduce emissions to the lowest achievable emission level
- Maximize recapture (recovery) and recycling
- Remove refrigerant before disposing of refrigerating systems
- Prohibit venting while installing, servicing, repairing, or disposing

On May 14, 1993, EPA issued final regulations on Section 608, which is discussed further in Chapter 9.

***Section 609 regulations*** — This section is called *Servicing Motor Vehicle Air Conditioners (MVACs)*. Section 609 requires that persons:

- servicing or repairing MVACs must properly use certified recycling equipment meeting SAE standards and EPA certification.
- servicing MVACs must be properly trained and certified by a certification program approved by EPA.
- buying refrigerant in less than 20-lb containers must be certified.

On July 14, 1992, EPA issued final regulations on Section 609, which are discussed in Chapter 10.

***The CFC tax*** — Congress levied an excise tax on the sale of CFCs and other chemicals that deplete the ozone layer. This was part of the Omnibus Budget Reconciliation Act of 1989. The tax was to progressively increase up to $4.95 per pound and to be collected on floor stocks of new refrigerant. Used refrigerant was exempted to encourage recycling.

# Energy Efficiency and Global Warming

While the ozone layer protects the earth from harmful UV-B radiation, certain "greenhouse gases" tend to block heat radiating from the Earth. Without greenhouse gases, the Earth would be considerably colder and uninhabitable by life as we know it.

Some scientists have become concerned that further increases in the concentration of greenhouse gases may cause a warming of the Earth's temperature. They believe this would cause a shift in climate patterns, melting of the polar ice caps, and would otherwise affect plant, animal, and human life.

Carbon dioxide is the most plentiful greenhouse gas. It is produced by burning fossil fuels at power plants and in automobiles. On the other hand, $CO_2$ is consumed by plants in the process called photosynthesis, whereby they "inhale" $CO_2$ and "exhale" oxygen. So, deforestation also increases the level of greenhouse gases in the atmosphere.

Air conditioners and refrigerating systems contribute to an indirect global warming effect by the electric power they consume. This, along with the electric power costs, leads to the emphasis on energy efficiency.

Most refrigerants have a global warming potential much higher than $CO_2$ due to their long atmospheric lifetime. When refrigerants are vented, they have a direct global warming effect. Leak-tight systems (containment) and recycling minimize the direct global warming effect in the same manner as they minimize ozone depletion. While some replacement refrigerants have zero ozone depletion potential (ODP), they have a significant direct global warming potential (GWP). *Therefore, containment and recycling are sound practices for all refrigerants.*

When considering the relative merit of substitute refrigerants, it is necessary to account for the indirect (energy efficiency) and the direct (GWP) effects. The total environmental warming impact (TEWI) factor is used to describe this total effect for various refrigerants.

## Impact of Environmental Concerns

The refrigeration industry has drastically changed since the signing of the Montreal Protocol in 1987. The refrigerants used in well over 95% of the world's systems will be phased out. Many who have spent a lifetime designing or servicing R-11, R-12, R-114, R-500, and R-502 systems, will see these systems replaced or retrofitted with an alphabet soup of new refrigerants. Soon, the R-22 phaseout will begin.

Who's affected by environmental concerns over refrigerants? It's actually easier to answer who isn't affected — nobody. The engineer, wholesaler, contractor, and mechanic are affected. So are the system manufacturer, chemical producer, and component manufacturer. Beyond that, the consumer and general public will be greatly affected.

Society and the general public will still be able to enjoy the comfort of a cool environment and the miracle of a refrigerated food supply. However, they will view these modern conveniences, or the refrigerants and refrigerating machines that produce them, with a new sense of appreciation and value.

## Refrigerant Life Cycle

Refrigerants are now viewed as renewable resources due to high prices, limited supply, increased public environmental awareness, and regulations. This new perspective changes the way and the cost of doing business.

While the rules and practices may be complex, the aims can be simply stated as follows:

- Select refrigerants for new or retrofit systems that minimize ozone depletion and global warming effects.
- Maximize energy efficiency in system design and operation.
- Contain refrigerants in systems and keep them out of the atmosphere.
- Use recycling practices that preserve the integrity of the used refrigerant supply.

Refrigerants will now be managed from "cradle to grave." New refrigerant will be charged into new equipment and will also make up for losses during operation. Used refrigerant will be isolated in the system or recovered for cleaning and reuse. Damaged or mixed refrigerant will be properly disposed of.

## Summary

Before the signing of the Montreal Protocol, refrigerants were viewed as inexpensive, harmless chemicals. Refrigerants were vented during service and new refrigerant was used to recharge systems. Ozone depletion and global warming are two environmental issues that have drastically changed this view.

The Montreal Protocol, signed in 1987, called for the phaseout of CFC refrigerants. The 1990 U.S. Clean Air Act Amendments called for phasing out and recycling of CFC and HCFC refrigerants. Refrigerants are now viewed as renewable resources due to their high cost and potential environmental damage.

All segments of the refrigeration industry are concentrating their efforts to 1) select new refrigerants; 2) maximize energy efficiency; 3) contain refrigerants; and 4) use recycling practices to preserve the used refrigerant supply.

Chapter Three

# Refrigerant Recovery Equipment

*Recover* means to remove refrigerant in any condition from a system and to store it in an external container. Refrigerant recovery can be accomplished in many ways, from simply evacuating or cooling the storage container, to using very complex automatic machines. Recovering can be understood by studying some of the functions of a vapor pumpout unit, as shown in Figure 3-1.

Each function of a recovery unit is interrelated and parallel to similar functions within a refrigeration system. In the first part of the chapter, we will discuss some of these functions and similar system components.

## Heat Exchangers

The evaporator boils liquid refrigerant so it can be pumped as a vapor by the compressor. Evaporator loading varies widely in refrigerant recovery applications. When recovering liquid refrigerant, the evaporator operates at full load until all the liquid is removed, then operates at no load for the final pumpdown into a vacuum. When recovering vapor refrigerant, the evaporator operates at no load throughout the recovery operation.

In recovery equipment, the useful work of the condenser provides heat for boiling refrigerant in the evaporator under full- or partial-load conditions. The excess heat is rejected to the atmosphere. The condenser loading also varies widely.

In early work, Cain[1] recognized the mutual benefit of combining the evaporator and condenser functions. As shown in Figure 3-2, a combination coil was used with separate condenser and evaporator circuits. The single fan blows first across the condenser section, then across the evaporator section. The condenser prewarms the ambient air, subsequently adding heat to the evaporator. (The direct metal contact between condenser and evaporator sections produces similar results.)

The Challenge of Recycling Refrigerants

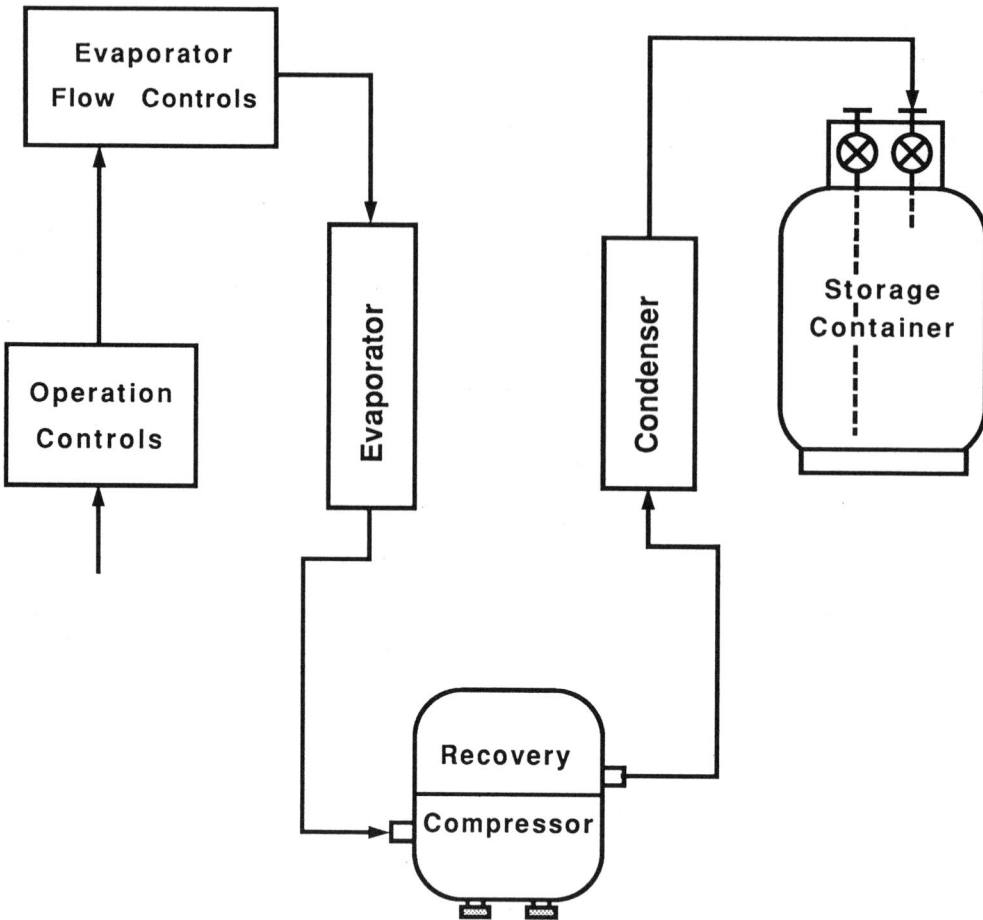

Figure 3-1. Recovery equipment functions

When the evaporator is in a no-load condition, the condenser functions as an air-cooled condenser but with additional mass and fin area. When using an air-cooled condenser, the usual choices concerning the number of rows, how many fins per inch, type of fins and tubes, and air velocity must be considered.

A combined heat exchanger and oil separator in a single canister[2] can be used in automotive recovery units, Figure 3-3. The evaporator inlet and outlet are located at the top of the canister with an oil drain located at the bottom. The condenser section, consisting of coiled tubes within the canister, also has a top inlet and outlet. The hot condensing refrigerant within the tubes rejects heat to the cool liquid or vapor surrounding the tubes. While the condenser heat rejection would have proved inadequate for sustained operation, the automotive charge is small and an interval is required to connect to the next car. This compact heat exchanger-oil separator has proven reliable and has been used in more automotive refrigerant recycling units than any other design.

# Refrigerant Recovery Equipment

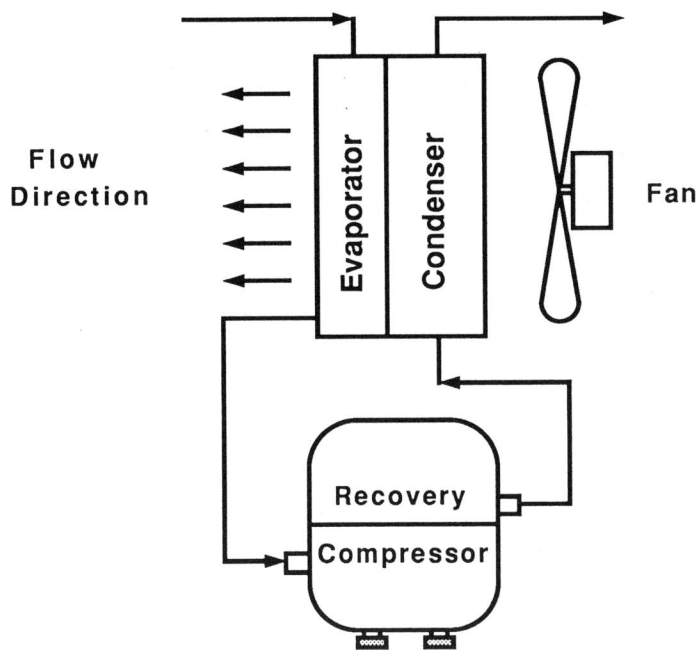

*Figure 3-2. Combination heat exchanger coil*

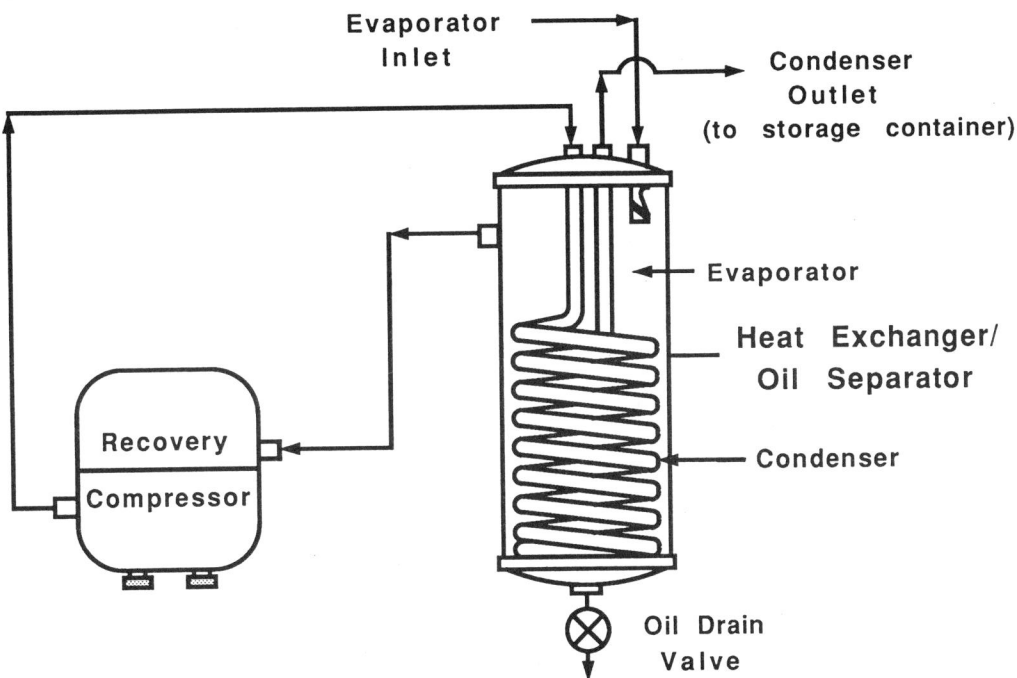

*Figure 3-3. Heat exchanger and oil separator for automotive recovery unit*

19

# The Challenge of Recycling Refrigerants

The coaxial coil shown in Figure 3-4 is another form of heat exchanger. Typically it is plumbed in a counterflow arrangement, so that the evaporator inlet and condenser outlet would be at the same end. To further increase capacity, it is possible to add coils around the compressor return oil separator, or add an additional air- or water-cooled condenser, Figure 3-5.

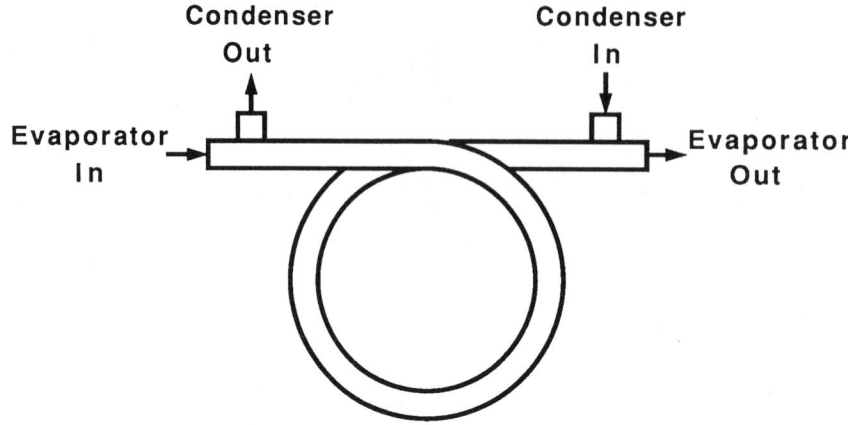

Figure 3-4. Coaxial coil heat exchanger

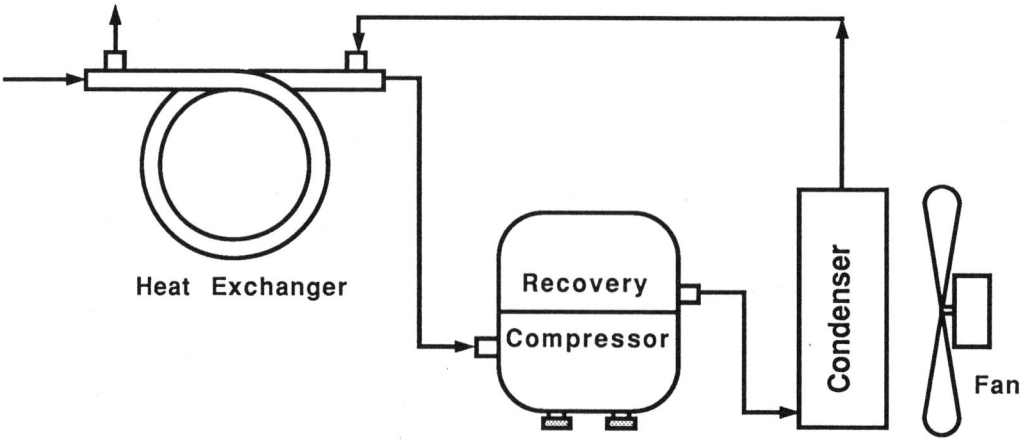

Figure 3-5. Heat exchanger with auxiliary condenser

## Compressors

The compressor lowers the suction pressure to the desired evaporating temperature, raises the discharge pressure to the required condensing temperature, and pumps the refrigerant into the storage container.

Compressor load varies. During most of the recovery cycle, the compression ratios and mass flow rates are similar to refrigeration system applications. During final vacuum operation, the compressor experiences a high compression ratio and low mass flow rate.

A typical compressor for an automotive recovery unit is shown in Figure 3-6. Originally designed with mineral oil for R-12 refrigerators, this compressor today comes charged with ester lubricant and is fully compatible with R-12 or R-134a.

Figure 3-6. Automotive recovery unit compressor (Courtesy, Americold Compressor Co.)

A larger commercial recovery compressor, Figure 3-7, was designed for R-502 but is compatible with many refrigerants, including R-12, R-134a, R-22, and R-500. An oil level sight glass is included.

Some of the compressor lubricant is discharged with the refrigerant in reciprocating and rotary compressors. An oil collection system, including a discharge oil separator[3], is shown in Figure 3-8. This oil separator preferably has a coalescing element that is 99%-plus efficient. A check valve is located at the discharge to prevent filling the separator with liquid refrigerant when the compressor shuts off.

The warming coils around the separator keep refrigerant from condensing during cold compressor starts and provide heat for boiling what does condense. The normally open oil return solenoid valve in this design closes when the compressor starts. The solenoid opens when the compressor shuts down to return oil to the compressor and to balance the load for easier starts. The oil return valve may be opened for oil return during long periods of operation.

# The Challenge of Recycling Refrigerants

*Figure 3-7. Commercial recovery compressor unit (Courtesy, Copeland Corporation)*

The compressor may also be used as a vacuum pump[4], as shown in Figure 3-9. After reaching the desired final vacuum, such as 4 in. (102 mm) mercury (Hg) for automotive air conditioners, the compressor discharge switches to atmosphere. This can take place at a later time, after repairs have been made.

Oil-less compressors, Figure 3-10, have been introduced in some recovery equipment. They offer the advantage of eliminating the need for discharge oil separators and associated oil return means. Also, not having an oil sump to dissolve refrigerant is an advantage; any contaminants tend to be pumped through.

The compressor load should still be balanced before start-up. Experience with these compressors is limited. When preceded by a very high quality suction oil separator, the oil-less compressor must operate without lubricant in the refrigerant. In recycling applications, the main oil separator should be placed after the oil-less compressor.

Under new technology, the thermal electric cooler[5] (TEC) shows promise in building a heat engine to replace the mechanical compressor for recovery-recycling applications. The cold side of the TEC draws liquid refrigerant into one compartment of the

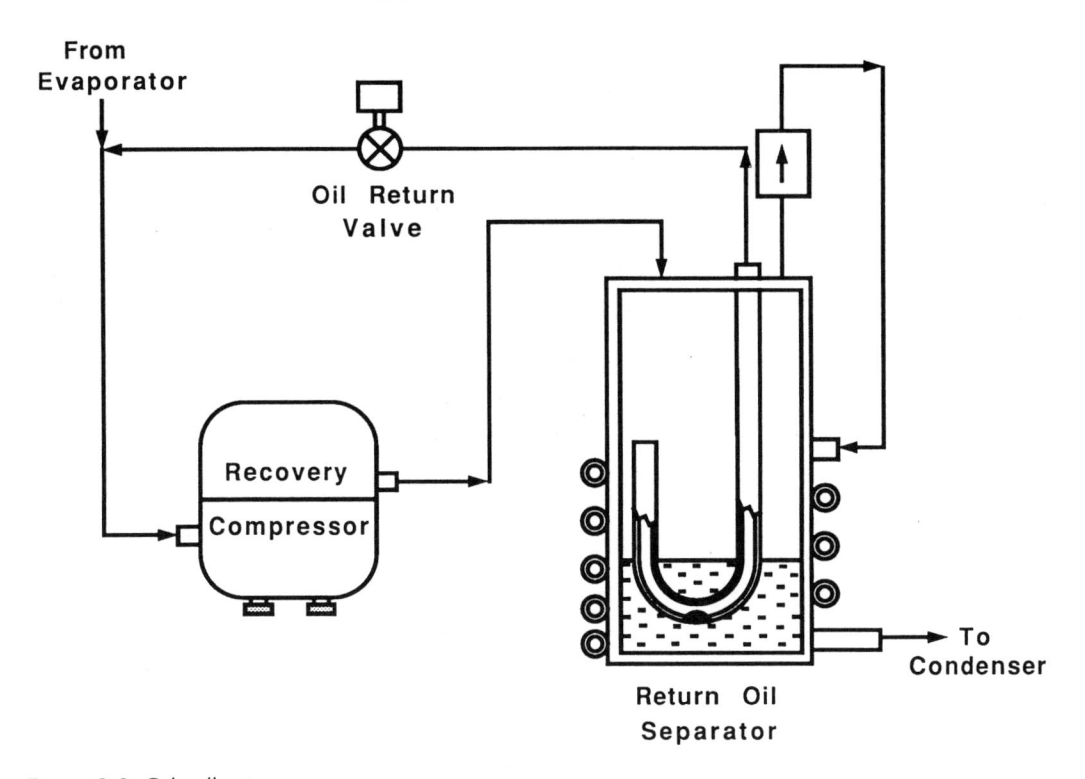

*Figure 3-8. Oil collection system*

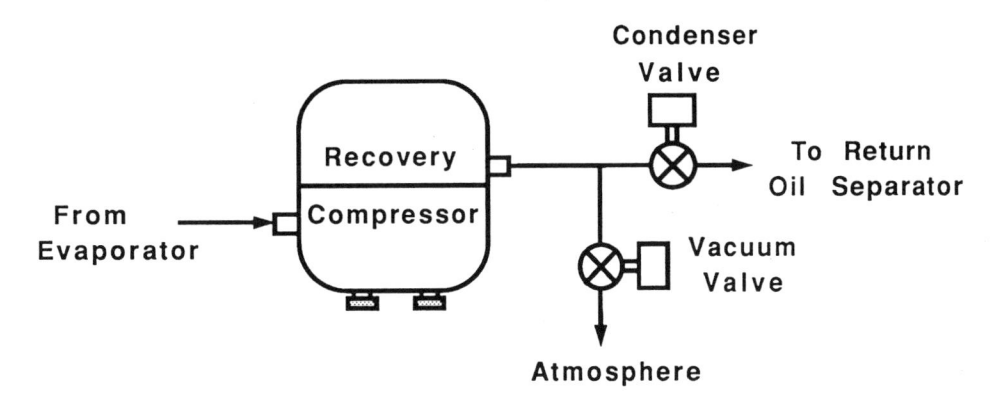

*Figure 3-9. Using a compressor as a vacuum pump*

two-compartment pump, while the hot side boils off refrigerant in the other compartment. Appropriate valves are used to fill and exhaust refrigerant as the two chambers alternate from cooling to boiling and to drain oil from the pump.

The inefficiency of the TEC is actually an advantage, as more heat is required in the boiling compartment. The pump boiling compartment is expected to operate well

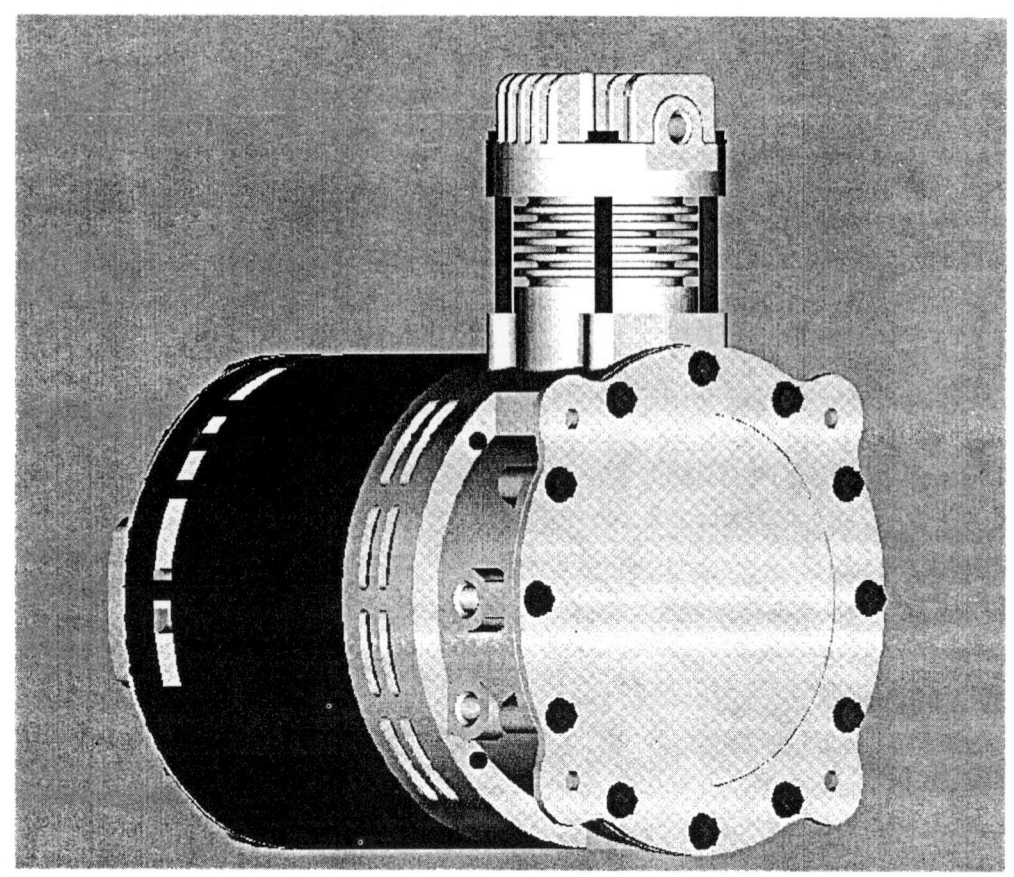

*Figure 3-10. Oil-less compressor (Courtesy, Thomas Industries Inc.)*

below ambient temperature, resulting in lower discharge pressures than mechanical compressor systems. Using appropriate finned surfaces, the pump boiling compartment may draw additional heat input from the ambient air, increasing pumping capacity.

## Storage Containers

The container stores used refrigerant and protects it from outside contamination. Most refrigerants are under pressure at ambient temperatures. Used refrigerants should never be stored in unapproved containers, such as the DOT (U.S. Department of Transportation) Specification 39 disposable cylinder shown in Figure 3-11.

Federal law forbids transporting DOT 39 cylinders if refilled; penalties up to $25,000 and five years of imprisonment may be expected [(Title 49 U.S.C. (United States Code) Sec. 1809)]. Proper storage containers, shown in Figure 3-12, are DOT-approved refillable containers stamped with the following information:

## Refrigerant Recovery Equipment

*Figure 3-11. Disposable refrigerant container (not suitable for storing used refrigerant)(Courtesy, Du Pont Company)*

    DOT-4BA400
    MANCHESTER D.B.
    W.C. 47.6 T.W. 28
    4-91
    "1st RETEST DATE - 96
    TEST EVERY 5 YRS
    THEREAFTER"

The first line is the DOT code for a tank rated at 400-psig (2,760-kPa) working pressure. The second line is the tank manufacturer's identification. W.C. is the water capacity in pounds when completely filled, and T.W. is the cylinder tare weight in pounds. (Since common refrigerants are heavier than water, this particular tank can safely be filled with 50 lb (23 kg) of most refrigerants.) The fourth line is the manufacturer's date code, which is important for the retest date.

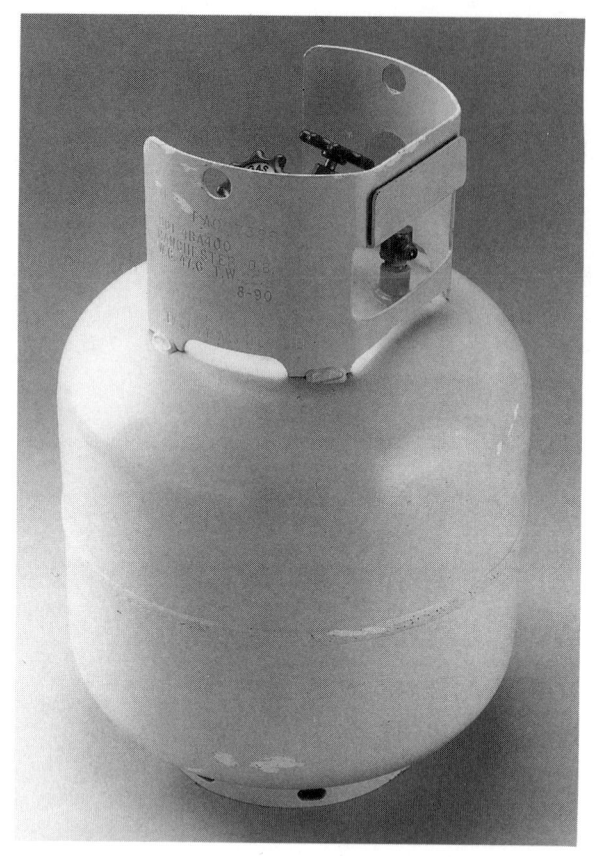

*Figure 3-12. Refillable container (suitable for storing used refrigerant) (Courtesy, Robinair Division, SPX Corporation)*

Containers and valves should meet UL Standard 1963[6], which requires the following:

- Storage container shall comply with DOT Specifications 49 CFR and have a service pressure rating not less than the recovery equipment's pressure limiting device.
- Cylinder valve shall comply with the Standard for Cylinder Valves, UL1769.
- Pressure relief device shall comply with the Pressure Relief Device Standard Part 1 — *Cylinders for Compressed Gases*, Compressed Gas Association Pamphlet S-1.1.
- Tank assembly shall be marked to indicate the first retest date, which shall be five years from date of manufacture. Also, the marking shall indicate that retest must be performed every subsequent five years. The marking shall be in letters at least 0.25 in. (6 mm) high.

# Refrigerant Recovery Equipment

ARI Guideline K[7] describes a color-coding scheme for refrigerant cylinders. The color for recovered and recycled refrigerants, as well as refrigerant on its way back to a reclaim facility, is gray with a yellow top. Some recycling equipment manufacturers have provided light-blue cylinders with yellow tops for automotive R-134a applications.

Most storage containers are equipped with SAE[8] 1/4-in. (6.4-mm) refrigeration flare fittings. For automotive R-134a applications, both new refrigerant cylinders (DOT 39) and refillable storage containers are fitted with special SAE[9] 1/2-in. (12.7-mm) threads.

## Container Labeling

EPA regulations[10] require labeling refrigerant tanks containing Class I (CFC) and Class II (HCFC) refrigerants. Each container shall bear the following warning statement:

*Warning: Contains [Insert refrigerant], a substance which harms public health and environment by destroying ozone in the upper atmosphere.*

An example is shown in Figure 3-13.

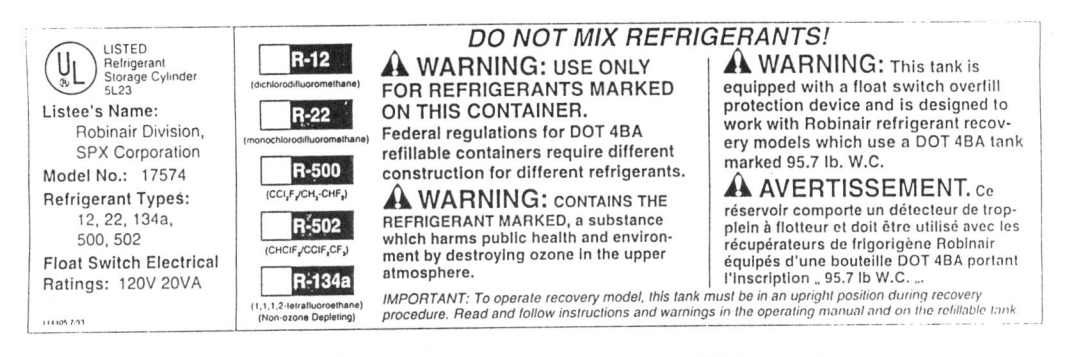

*Figure 3-13. Storage container label (Courtesy, Robinair Division, SPX Corporation)*

## Filling Containers

Care must be exercised when filling storage containers, because liquid refrigerants expand when heated. Referring to Figure 3-14, the tank on the left is filled to an 80% level at 70°F (21°C). The tank on the right shows that same amount of refrigerant warmed to 158°F (70°C).

When a container is overfilled and subsequently exposed to higher temperatures, hydraulic expansion can cause the relief valve to open and vent refrigerant, or it can rupture the container. SAE[11,12,13] and UL[6] standards require container overfill protec-

**80% Full at
70°F (21°C)**

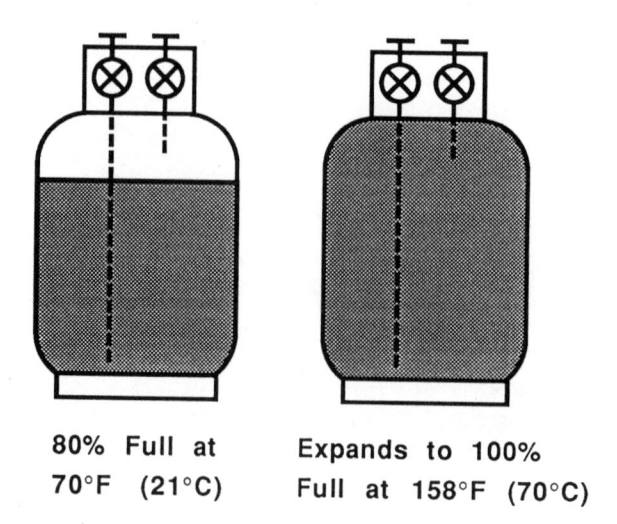

**Expands to 100%
Full at 158°F (70°C)**

*Figure 3-14. Expansion of liquid refrigerant when heated*

tion based upon 80% liquid fill at 70°F. A full container condition can be determined by monitoring weight while filling. A method using a load cell[14] is shown in Figure 3-15.

A predetermined load deflects the load cell so as to trip a limit switch. An electronic scale used for charging can also detect a tank overfill condition, Figure 3-16. For recovery units handling more than one refrigerant, allowable weight limits must be based on the refrigerant with the lowest density to prevent overfilling the container.

Float switches, or other liquid-level sensors, can be used to detect full containers. They correctly gauge the fill level regardless of the refrigerant type or oil content. However, they add to the cost of each container, versus adding the cost of a load cell to the recovery unit, which allows using standard containers.

## Retesting Cylinders

DOT regulations[15] require that cylinders be retested every five years. The appropriate categories are DOT-3A, DOT-3AA, DOT-3A480X, DOT-4B, DOT-4BA, DOT-4BW, and DOT-4E containing fluorinated hydrocarbons and mixtures thereof, which are commercially free from corroding components. Two retesting options are available:

- External visual examination with hydrostatic test
  — Retest at five-year intervals.
  — An authorized retester must perform a visual examination using CGA pamphlet C-61.
  — Conduct a hydrostatic test at two times the service pressure.
  — Measure expansion during hydrostatic test.

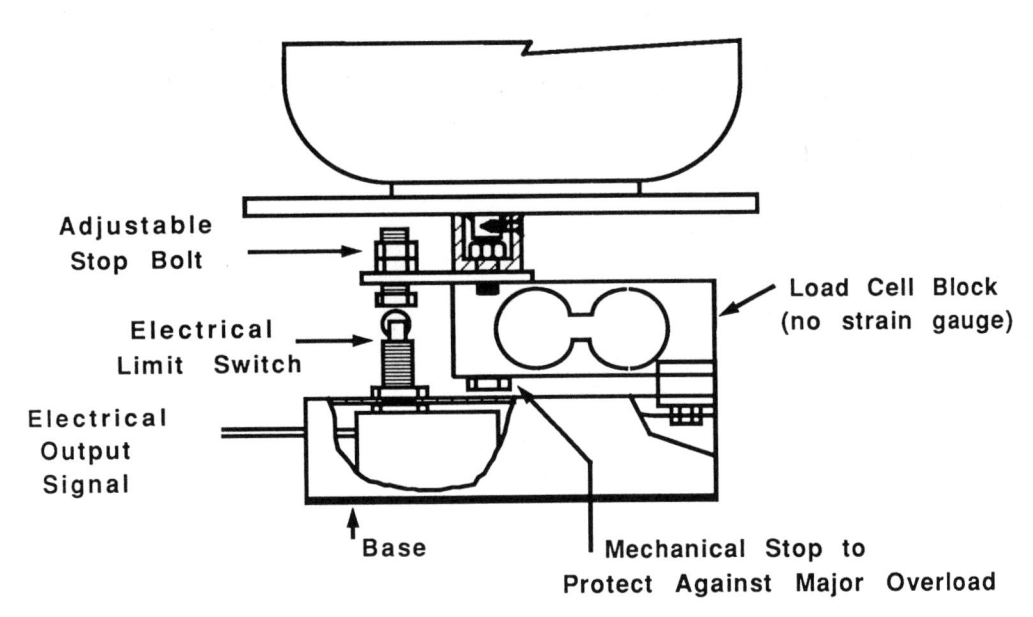

*Figure 3-15. Mechanical-electrical load cell tank overfill protection*

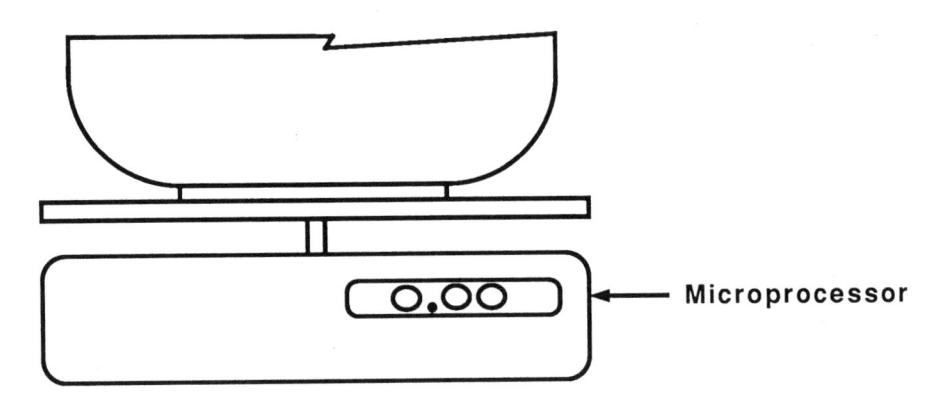

*Figure 3-16. Electronic scale tank overfill protection*

— Condemn container when tank leaks, has external corrosion, denting, bulging, or evidence of rough usage exists, or when the permanent expansion exceeds 10% of the total expansion after the hydrostatic test.
— Owner must keep records showing the result of reinspection and retesting until the cylinder is again reinspected or retested.
— Mark each cylinder passing retest with the cylinder retester's identification number set in a square pattern, between the month and year of the retest date in characters not less than 1/8 in. (3 mm) high, with the first character occupying the upper left corner of the square pattern. For example, a cylinder retested in May, 1984 by approved retester, A123 would be stamped:

```
        A   1
   5            84
        3   2
```

☐ External visual examination only
  — Retest at five-year intervals.
  — Performed by competent person.
  — May not apply where substantial amounts of used oil have been introduced (non-corrosive).
  — Perform a complete external visual examination per CGA Pamphlet C-6.
  — Owner must keep records including date of inspection (month and year); DOT specification number; cylinder identification (registered symbol number and serial number, date of manufacture); type of cylinder protective coating (painted, etc., and statement as to need for refinishing or recoating); conditions checked (leakage, corrosion, gouges, dents or digs in shell or heads, broken or damaged footring or protective ring, or fire damage); disposition of cylinders (returned to service, to cylinder manufacturer for repairs, or scrapped).
  — Mark cylinder by date (month and year) followed by the letter "E." For example, a cylinder retested by visual external inspection in May 1984 would be marked 0584E.

## Evaporator Flow Controls

Flow controls meter refrigerant to the evaporator to prevent slugging the compressor with liquid. Since liquid or vapor refrigerant may be withdrawn from the system, the recovery unit must function properly for both. The recovery unit must also process different refrigerants.

Capillary tubes[16] and thermostatic expansion valves (TXVs) have been used to throttle refrigeration systems. These systems are designed for use with a single refrigerant in which subcooled liquid is supplied to the expansion device inlet. The compressor, evaporator, and condenser are all selected to optimize performance for that refrigerant.

Recovery units for automotive air conditioners have been designed for handling either R-12 or R-134a, but not both. The refrigerants usually are recovered as a vapor, and the unit's internal accumulator can hold the entire 2- to 4-lb (0.9- to 1.8-kg) charge. Vapor recovery speed is emphasized, because flow control is not required.

In most stationary applications, it is impractical to provide an internal accumulator large enough to hold the maximum refrigerant charge. Compromises must be made when sizing recovery unit heat exchangers, compressors, and associated piping to handle multiple refrigerants. The design goal is to optimize speed, whether recovering liquid or vapor refrigerant.

### Capillary Tubes

A capillary tube can meter different refrigerants. However, capillary tube size and length requirements are quite different for each refrigerant. Sizing a capillary tube properly for one refrigerant may cause starving for a second refrigerant and flooding for a third.

The capillary tube cannot provide precise flow control over a broad temperature range as a TXV can. Capillary tubes can greatly impede vapor refrigerant flow.

The capillary tube system can be simply modified, Figure 3-17, by adding a three-way valve and a sight glass. The valve has separate vapor and liquid outlet ports. Refrigerant flows without restriction through the vapor port to the evaporator. In the liquid port position, refrigerant flows through the capillary tube on the way to the evaporator.

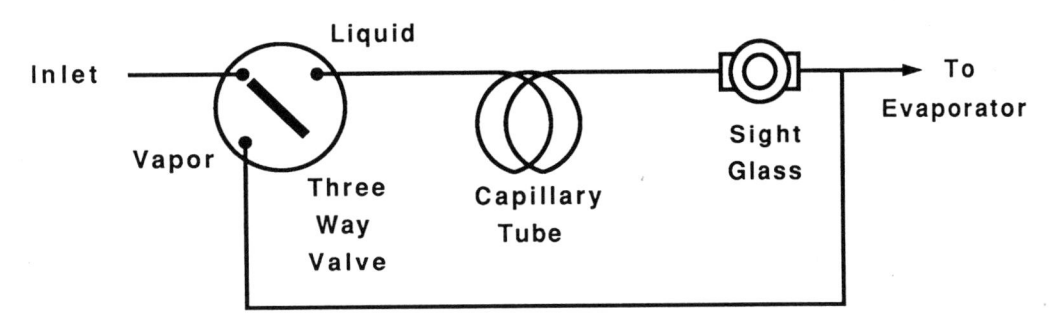

*Figure 3-17. Modified capillary tube flow control system*

The technician starts recovering through the liquid port. When the sight glass bubbles and clears, the technician immediately turns the valve to the vapor port to complete the recovery operation.

While operated manually, the modified capillary tube apparatus provides improved vapor-recovery speed. A capillary tube can be selected to provide reasonable flow control for R-12, R-22, and R-502. Replacing the sight glass with either a liquid sensor or a pressure switch and controlling the valve with a solenoid, provides further improvement.

## TXV

A TXV provides excellent liquid flow control for a single refrigerant. It fully utilizes evaporator capacity and controls superheat at the evaporator outlet over a broad temperature range.

TXVs meter refrigerant based on the difference in actual pressure and the pressure in a sealed bulb located at the evaporator outlet. The same refrigerant flowing through the evaporator is used in the sealed bulb. An internal spring sets the superheat value. For the refrigerants under consideration, the TXV is a one-refrigerant control.

The TXV diaphragm movement limits valve travel. For this reason, vapor refrigerant flow is restricted even in the fully open valve position.

Van Steenburgh[17] has used a parallel path for each refrigerant with a solenoid-operated valve and a TXV in each path. A rotary selector switch with marked positions for each refrigerant controls the solenoid valves to provide the desired path. In recirculating operation, he always provides liquid refrigerant to the selected TXV.

A multiple-refrigerant TXV (MRTXV) was developed[17] in cooperation with the Sporlan Valve Co., Figure 3-18. This MRTXV includes two diaphragms and two sealed bulbs with the same refrigerant or powered element in each bulb. The first bulb is placed in the normal TXV bulb location at the evaporator outlet. The second, or unique bulb, is placed at the evaporator inlet.

The MRTXV meters refrigerant flow to provide constant temperature rise across the evaporator for any refrigerant, much the same way a TXV meters for a single refrigerant. A single MRTXV could be stocked for aftermarket applications for a given tonnage rating. It could also be used on new refrigeration equipment when a retrofit is likely during the system lifetime.

Even when fully open, the MRTXV still limits vapor flow rate. This is in spite of maximizing valve travel, tapering the valve seat, and adding a small bleed hole. Combining the MRTXV with a low-side float arrangement[18], Figure 3-19, overcomes this problem. As long as liquid is at the float level, refrigerant is only fed to the MRTXV. When liquid falls below the float level, the vapor-recovery solenoid valve opens. Then refrigerant bypasses the MRTXV and the evaporator to minimize recovery time.

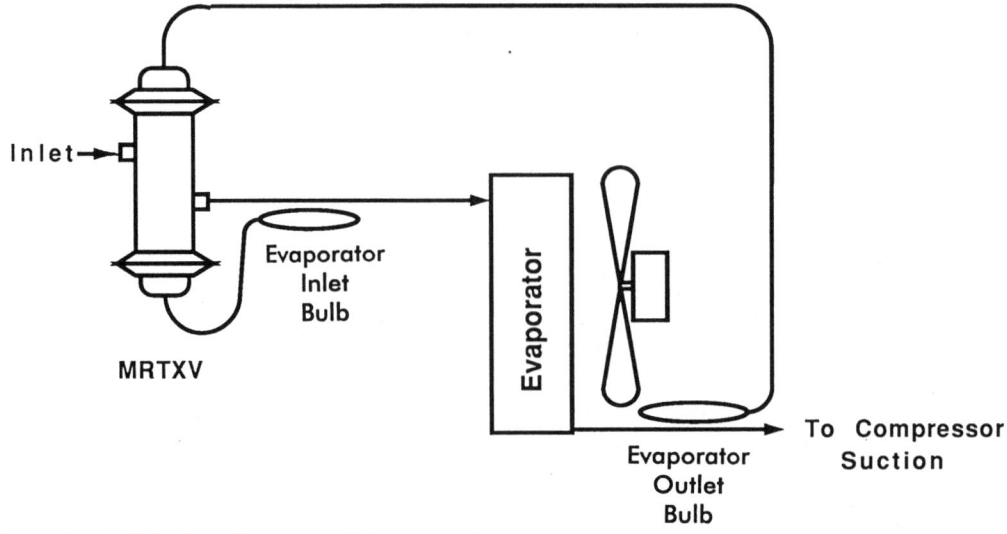

Figure 3-18. Multiple-refrigerant TXV flow control system

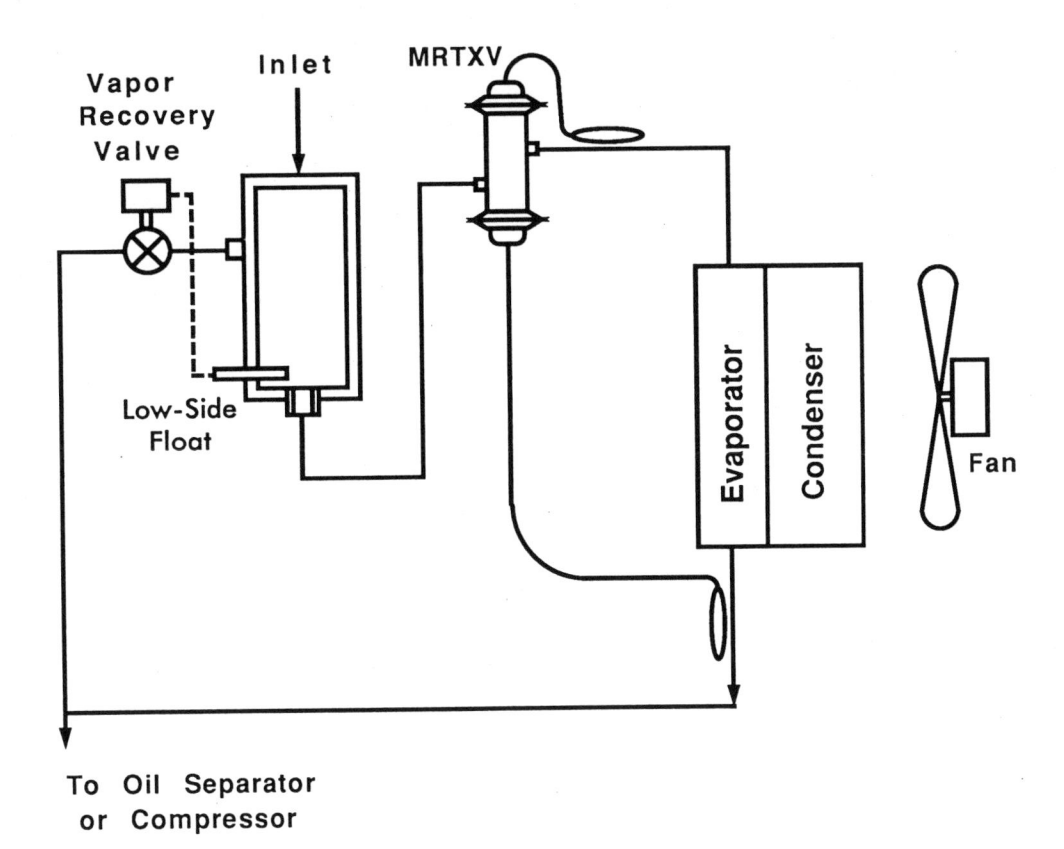

Figure 3-19. MRTXV with low-side float flow control system

## Low-Side Float

A low-side float provides excellent evaporator control when using the heat exchanger-oil separator of Figure 3-3. The condenser coils provide heat to boil off refrigerant; the float holds a liquid level above the top of the coils. This arrangement is preferred in recycling applications where liquid refrigerant is normally handled. It will be discussed in more detail later.

## Operation Controls

In early work, Cain[1] used a low-pressure switch in the suction line and a high-pressure switch in the discharge line to control the recovery compressor. He also used manual valves. Early automatically controlled prototypes[19] used many solenoid valves and both collection and new supply containers.

Simple fully automatic recovery controls[20] use a low-pressure switch, high-pressure switch, tank-full switch, and recovery solenoid valve. The schematic is shown in Figure 3-20; the electrical controls are shown in Figure 3-21.

Recovery operation initiates by pushing an electrical switch. The recovery solenoid opens and the compressor starts. Normally, the recovery operation automatically terminates by closing the recovery solenoid and shutting off the compressor when the low-pressure switch senses a predetermined vacuum.

When either the high-pressure switch senses an excessive pressure or the tank-limit switch senses an overfill condition, the recovery operation also automatically terminates. This remains the standard for simple automatic controls.

## Final Recovery Vacuum

The U.S. EPA[21,22] has set limits for the final recovery vacuum recovery units must achieve for the following applications:

- Small appliances, such as refrigerators and freezers must be capable of removing 90% of the charge if the refrigerator compressor is operating and 80% if it is not.
- Automotive air conditioners and mobile-like a/c (farm, construction equipment) — 4-in. Hg (14-kPa) vacuum.

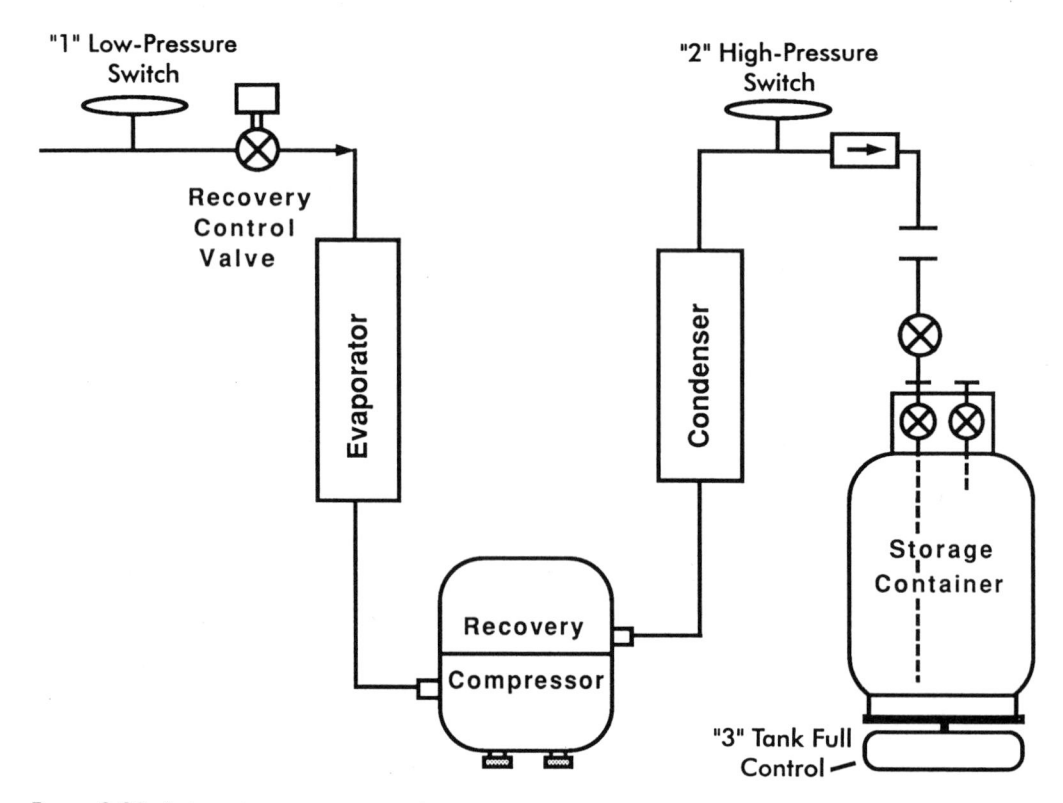

*Figure 3-20. Automatic recovery controls*

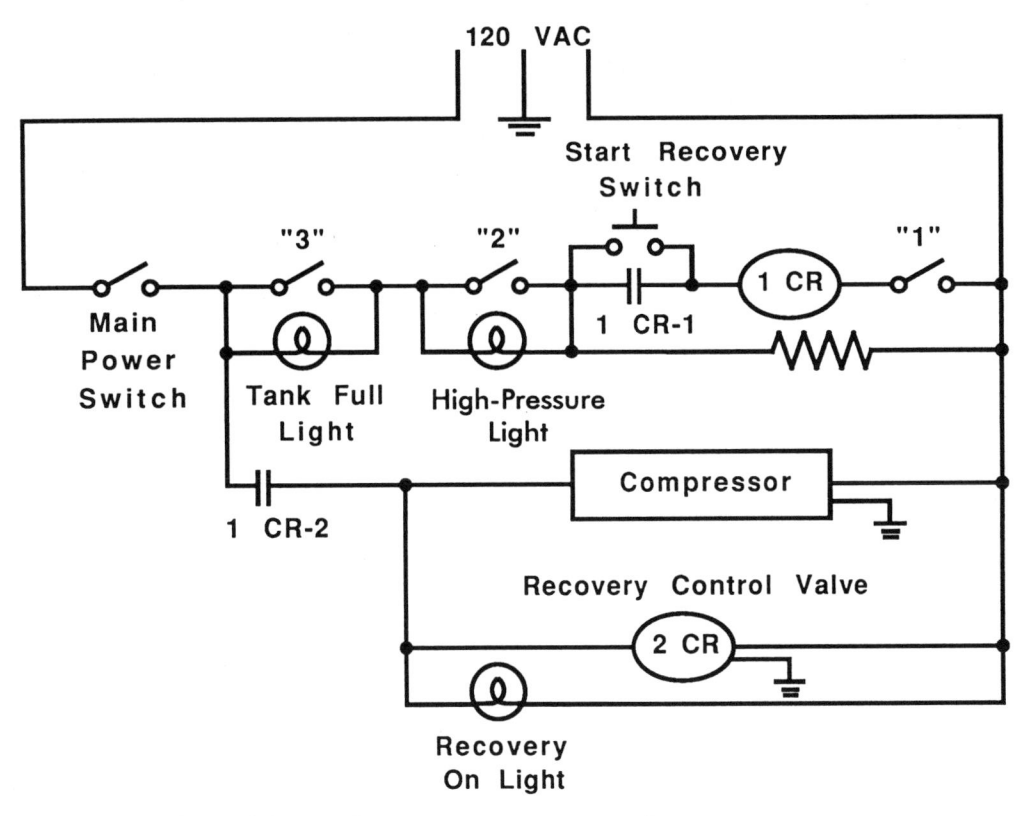

*Figure 3-21. Electrical diagram for automatic recovery controls*

- High-pressure systems (R-12, R-502) up to 200-lb (91-kg) charge and all R-22 systems — 10-in. Hg (34-kPa) vacuum.
- High-pressure systems, except R-22, over 200-lb (91-kg) charge — 15-in. Hg (51-kPa) vacuum.
- Low-pressure systems — 29-in. Hg (91-kPa) vacuum.

EPA requires certification of recovery units to meet their requirements. Currently, the Air-Conditioning and Refrigeration Institute (ARI) and Underwriters Laboratories (UL) have been approved by EPA for this purpose. They use ARI Standard 740-93[23] methods.

## Clearing Recovery Equipment

When the recovery operation ends, the system and the recovery unit low side are in a vacuum. The recovery unit high side is at storage container pressure and contains liquid refrigerant. Since refrigerant mixing causes serious problems (see Chapter 7), refrigerant trapped in the condenser[16] must be removed to avoid contamination with the next refrigerant.

## Make-Shift Methods

One method of removing trapped refrigerant is to gently heat the condenser section with an electric heat gun (blow dryer). Another is to cool the storage container to collect most of the refrigerant. A third method is to use an auxiliary recovery unit, wherein the refrigerant is mixed with other refrigerants. The resulting quantity must be disposed of but is small compared to the total refrigerant recovered.

A fourth method is to provide a valved fitting near the condenser of the recovery unit. My son, Tony, used the method shown in Figure 3-22, for an eighth grade science project report. He evacuated the container and connected a specially configured metal tube to the valved fitting with a specially configured automatic shutoff hose[24]. The metal tube was placed in an ice bath, so as to cool the refrigerant in the metal tube before it passed into the storage container.

Placing the metal tube in the ice bath showed improved results over placing the storage container in the ice bath. He further removed refrigerant by repeating the procedure into a second evacuated container. The average amount of trapped refrigerant recovered was 54.1% with one evacuated tank and 80.1% with two successive tanks.

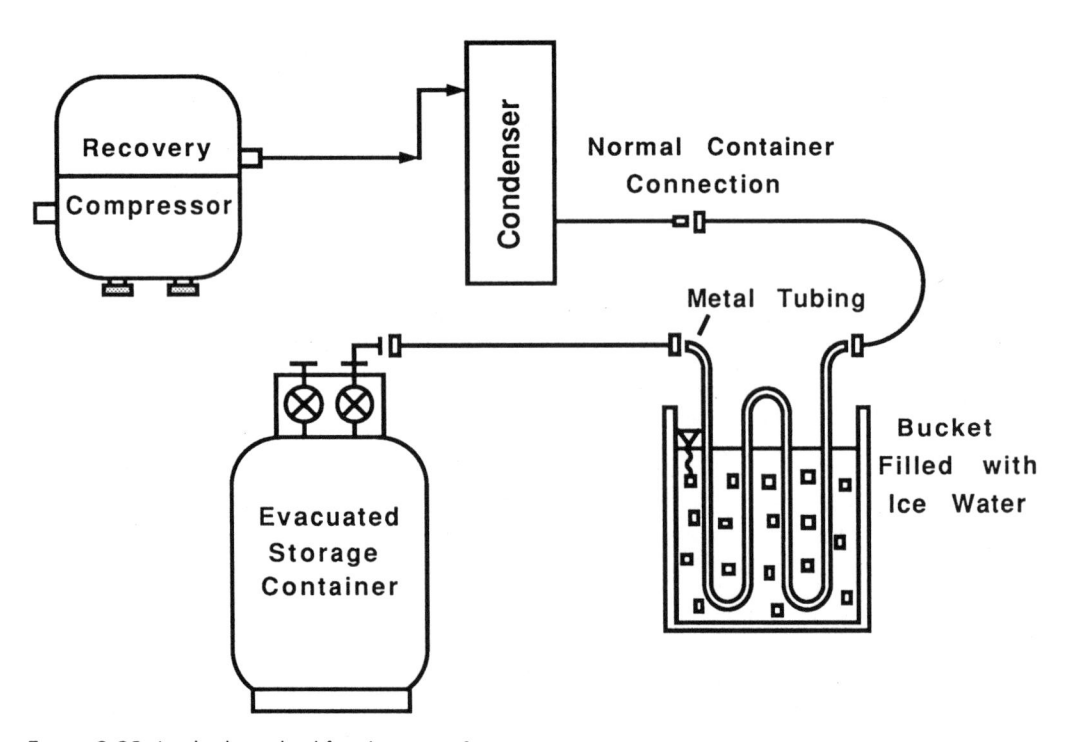

*Figure 3-22. Ice bath method for clearing refrigerants*

## Multi-Section Condenser

A condenser divided into multiple sections[25] can be used when portability and weight are not required. As shown in Figure 3-23, each refrigerant has its own section and valving. Refrigerant need not be cleared from the valved sections.

The small quantity of vapor left between the compressor and the separate valves is minimized with short lengths of piping. A service port provides for evacuating the trapped vapor using a vacuum pump.

## Clearing Valve

Laukhuf[26] and Murray cleared recovery equipment by reconnecting one hose so that the compressor discharge was directed to the storage container and a second hose so that the condenser was connected to the recovery inlet. For the brief time required to clear the recovery unit, the storage container served as the condenser.

An internal valving arrangement[27] using either two three-way valves or a single four-way valve accomplishes the same function. The currently preferred design uses a

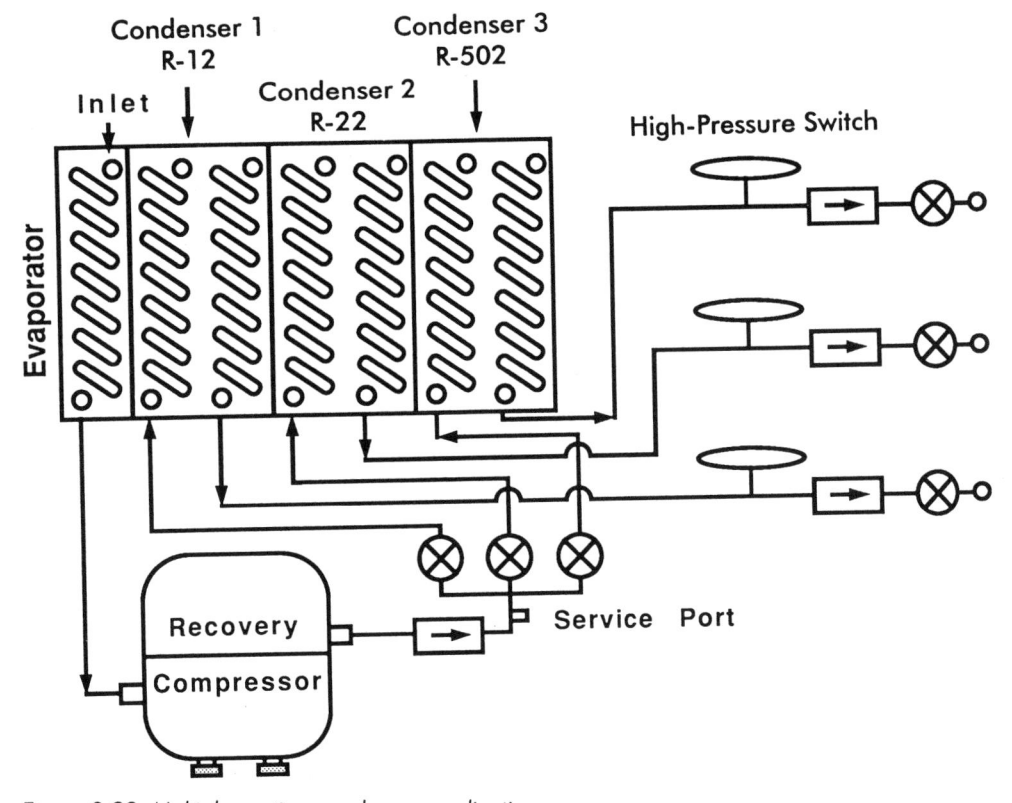

*Figure 3-23. Multiple-section condenser application*

four-way reversing solenoid valve originally designed for heat pump applications, Figure 3-24.

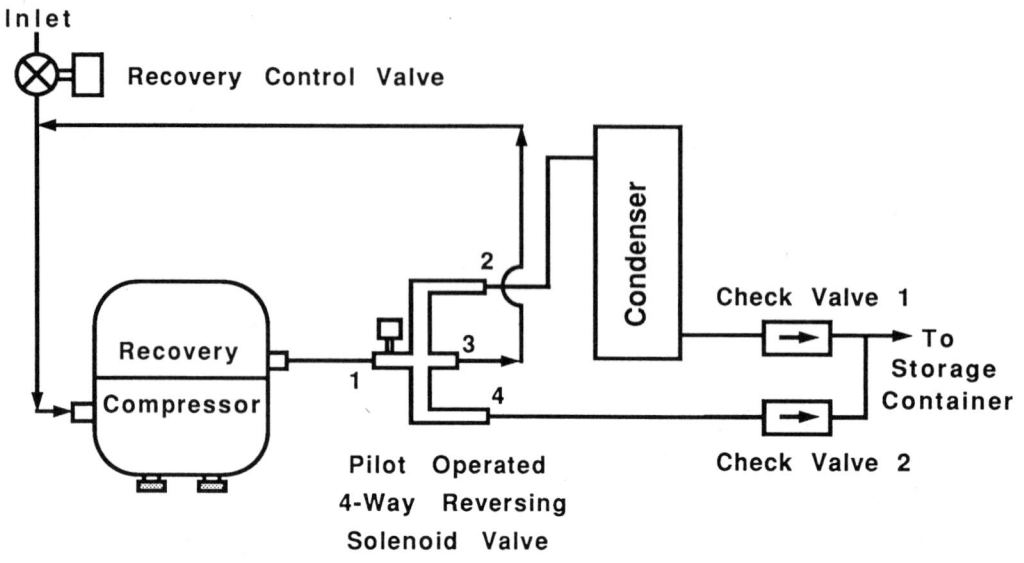

Figure 3-24. Clearing with four-way valve

In recovery mode, compressor discharge port 1 connects through port 2 to the condenser inlet, and then to the storage container through check valve 1. Meanwhile, port 3 to the recovery inlet is connected through port 4, so that the suction line extends to check valve 2.

In clearing mode, compressor discharge port 1 is connected to the storage container through port 4 and check valve 2. At the same time, port 3 is connected through port 2 to the condenser, so that the suction line extends to check valve 1.

When the unit is cleared to the low-pressure-switch setting, the operator-initiated clearing operation ends. A vacuum pump should then be connected to the recovery unit to completely clear the unit.

# Automotive Recovery Equipment

While recovery-recycle units are preferred by most motor vehicle manufacturers and service technicians, SAE J2209[12] covers extraction-only equipment for R-12 systems. These recovery units, such as the one shown in Figure 3-25, are preferred for collecting refrigerants from remote locations (farms) and at automotive salvage yards.

# Refrigerant Recovery Equipment

*Figure 3-25. Automotive recovery unit (Courtesy, Robinair Division, SPX Corporation)*

Units must be certified to meet SAE safety, performance, and feature requirements. They must meet the same recovery vacuum (4 in. Hg) and oil-measuring specifications as recycling equipment rated under SAE J1990. Vehicle manufacturers and other members of the SAE Defrost and Interior Climate Control Committee were concerned that the collected refrigerant would be used without proper cleaning. They prefer that such refrigerant be recycled on-site or be sent to a reclaim facility for cleaning and analysis. The following are special requirements unique to R-12 extraction equipment:

- Storage container must be marked in letters at least 0.8 in. (20 mm) high, "DIRTY R-12 — DO NOT USE, MUST BE REPROCESSED."
- Storage container shall have a 3/16-in. (4.8-mm) flare male thread connection as identified in SAE J639, *CFC-12 High Pressure Charging Valve* (Figure 3-2).
- Service hoses must be furnished meeting SAE J2196 refrigerant permeation specifications and having shutoff valves within 12 in. (30 cm) of the end of the hose.

## Other Recovery Methods

Not all possible methods will be considered. Substantial amounts of used oil are transferred to the storage container in the following methods.

### Cooling the Storage Container

When a separate refrigeration unit cools the storage tank, the recovered refrigerant doesn't come into contact with the refrigerant in the cooling unit. The separate unit evaporator is in heat exchange relation with the storage container, or a heat exchanger located before the storage container. The efficiency of such a recovery system depends entirely on the evaporator temperature of the cooling unit.

Scuderi[28] used a recovery compressor that was not located in the recovery path, but rather used the storage container as an evaporator. He drew vapor off the top of the container with the compressor and pumped it through an air-cooled condenser and a special pressure-reducing valve before dumping it back into the recovery path. In this manner, he was cooling the tank rather than pumping the refrigerant.

## Push-Pull Method

A vacuum pump is used for R-11 systems as shown in Figure 3-26. The suction port connects to the storage container, and the discharge port connects to the system vapor space. Separate piping connects liquid refrigerant in the system to the storage container. The higher pressure in the system pushes and the lower pressure in the container pulls liquid refrigerant into the storage container.

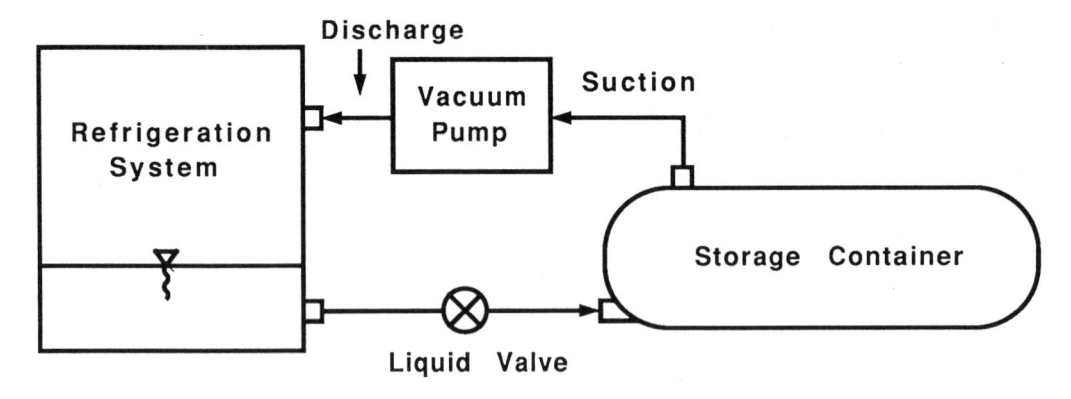

*Figure 3-26. Low-pressure refrigerant recovery unit*

For low-pressure refrigerants, the vacuum pump reverses to complete the pumpout when the liquid valve is closed. For high-pressure refrigerants, any vapor pumpout recovery unit may be connected first as a push-pull for liquid transfer and then as a vapor pumpout unit for full recovery. The connections are shown in Figure 3-27. On the left, liquid transfer connections are similar to the vacuum pump in Figure 3-26. On the right, vapor pumpout uses the normal connections.

A method with internal connections is shown in Figure 3-28. An optical sensor is used to automatically reconnect from liquid transfer to vapor pumpout. When liquid is sensed, the compressor draws vapor off the tank through valve L2, pumps it through the condenser and valve L3, and to the refrigeration system vapor port. The liquid flows by push-pull through valve L1 to the storage container. When vapor is

# Refrigerant Recovery Equipment

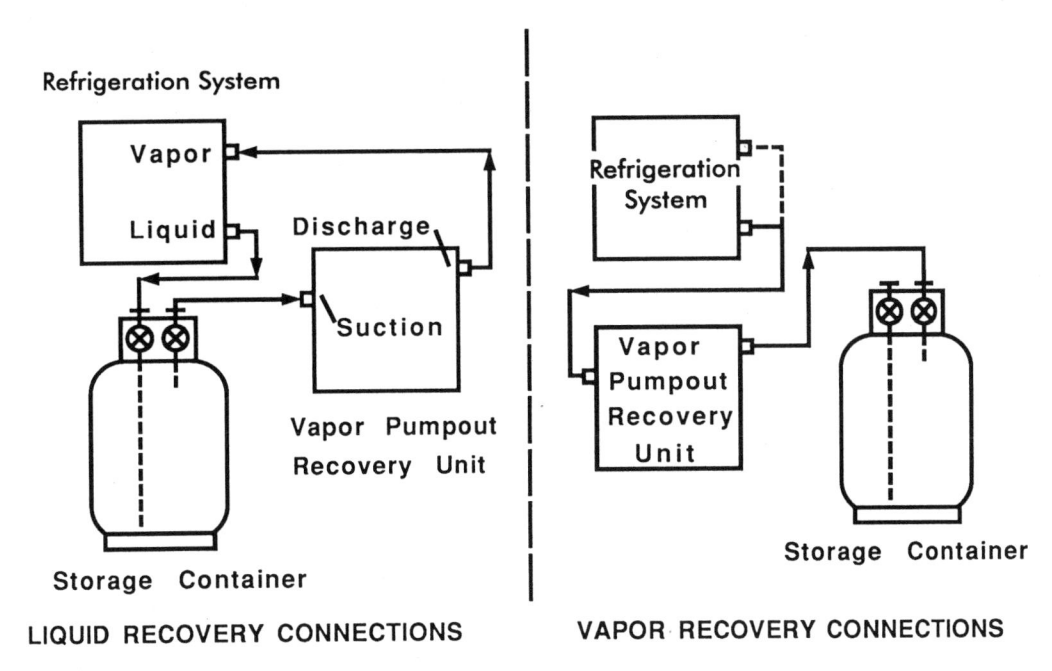

*Figure 3-27. Using recovery unit for push-pull liquid transfer*

sensed, the compressor draws vapor from the system through valve V1, pumps it through the condenser and valve V2, and to the storage container.

## Pumping Liquid

A liquid pump, such as the one shown in Figure 3-29, can directly transfer liquid refrigerant to a storage container. This applies to either low-pressure refrigerants like R-11 or to high-pressure systems with large quantities of refrigerant. Liquid pumps use displacement, such as gear-type construction. They do not have the ability to pump out vapor to federal requirements.

# Summary

Recover means to remove refrigerant in any condition from a system and to store it in an external container. Evaporators and condensers for vapor pumpout units are often in heat exchange contact.

Typically, standard compressors used in air conditioning and refrigeration systems are selected for recovery units. Storage containers must meet federal requirements, including retesting every five years. Special precautions in filling containers are required.

The Challenge of Recycling Refrigerants

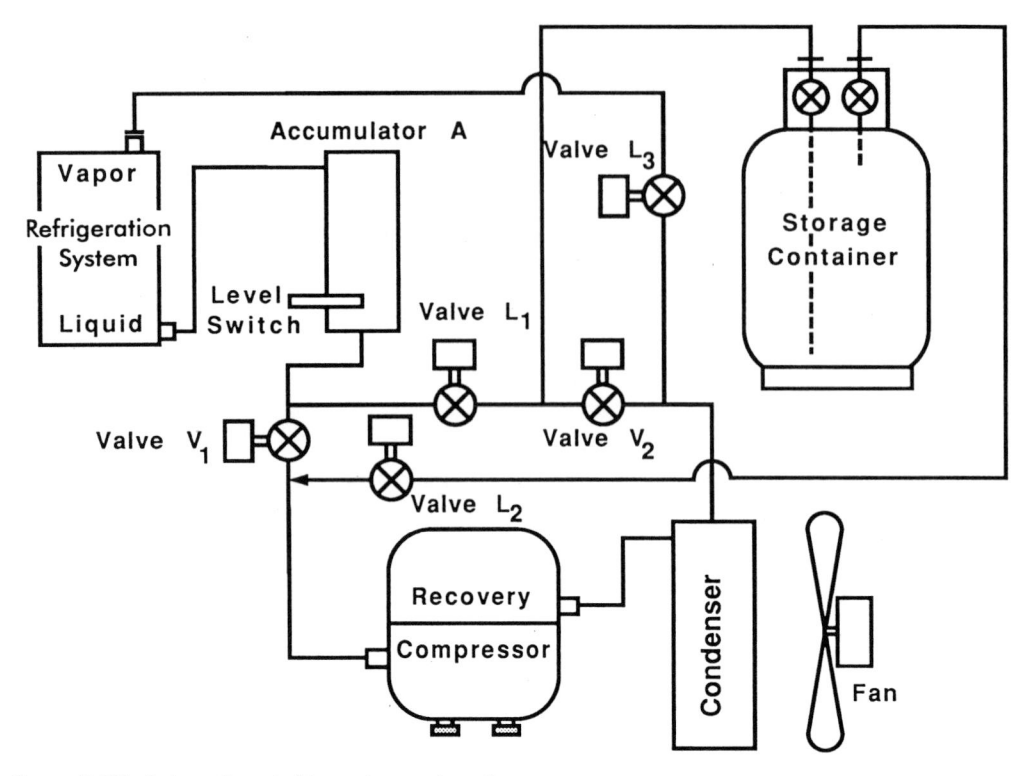

*Figure 3-28. Automatic switching using push-pull*

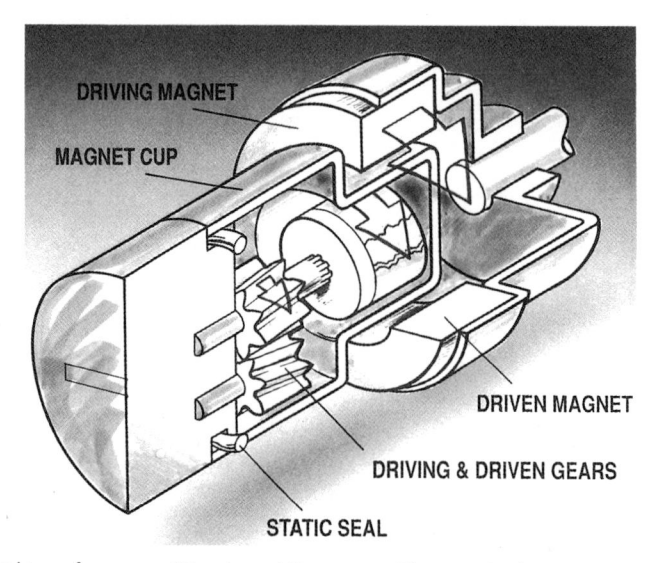

*Figure 3-29. Liquid transfer pump (Courtesy, Micropump Corporation)*

42

Evaporator flow controls must be specially designed for recovery units handling multiple refrigerants. Operation controls are used to terminate the recovery cycle upon achieving the final recovery vacuum and to protect against unit over-pressure or full-storage container conditions. Recovery units handling multiple refrigerants must be cleared of one refrigerant before switching to another.

Automotive extraction-only equipment and alternative methods to the vapor pumpout unit are suitable for many applications.

# References

[1] U.S. Patent 4,261,178, Robinair Division, SPX Corporation, 1981.

[2] U.S. Patent 4,809,520, Robinair Division, SPX Corporation, 1989.

[3] U.S. Patent 5,042,271, Robinair Division, SPX Corporation, 1991.

[4] U.S. Patent 4,805,416, Robinair Division, SPX Corporation, 1989.

[5] U.S. Patent Pending, Robinair Division, SPX Corporation.

[6] Underwriters Laboratories, UL 1963, First Edition, *Standard for Refrigerant Recovery-Recycling Equipment*, 1993, Copyright 1989.

[7] ARI Guideline K, 1990, *Containers for Recovered Fluorocarbon Refrigerants*.

[8] SAE J513, 1976, *Refrigeration Tube Fittings*.

[9] SAE J2197, 1991, *HFC-134a Service Hose Fittings*.

[10] Part III Environmental Protection Agency, 40CFR Part 82, *Protection of Stratospheric Ozone: Labeling Supplemental Proposal Final Rule.*

[11] SAE J1990, 1989, *Extraction and Recycle Equipment for Mobile Automotive Air-Conditioning Systems.*

[12] SAE J2209, 1992, *Extraction Equipment for Mobile Automotive Air-Conditioning Systems.*

[13] SAE J2210, 1991, *HFC-134a Recycling Equipment for Mobile Air-Conditioning Systems.*

[14] U.S. Patent 4,878,356, Robinair Division, SPX Corporation, 1989.

[15] DOT 49CFR Chapter 1 (10-1-92 Edition), Sections 173.34 (e) and 173.3 (d).

[16] Kenneth Manz, "How to Handle Multiple Refrigerants in Recovery and Recycling Equipment," *ASHRAE Journal*, April 1991.

[17] U.S. Patent 5,231,842, Robinair Division, SPX Corporation, 1993.

[18] U.S. Patent 5,203,177, Robinair Division, SPX Corporation, 1993.

[19] U.S. Patent 4,441,330, Robinair Division, SPX Corporation, 1984.

[20] U.S. Patent 4,768,347, Robinair Division, SPX Corporation, 1988.

[21] U.S. Environmental Protection Agency, Section 609, 1992.

[22] U.S. Environmental Protection Agency, Section 608, 1993.

[23] ARI Standard 740-93, 1993, *For Performance of Refrigerant Recovery and/or Recycling Equipment*, Copyright 1993.

[24] U.S. Patent 5,005,375, Robinair Division, SPX Corporation, 1991.

[25] U.S. Patent 4,939,905, Robinair Division, SPX Corporation, 1990.

[26] U.S. Patent 5,095,713, Robinair Division, SPX Corporation, 1992.

[27] U.S. Patent 5,127,239, Robinair Division, SPX Corporation, 1992.

[28] U.S. Patent 4,766,733, Carmelo J. Scuderi, 1988.

Chapter Four

# Oils, Acids, and Particulates

Refrigerant oil is used to lubricate the system compressor. Traditionally, mineral oils and alkylbenzenes have been used with the CFC and HCFC refrigerants with which we are familiar. New refrigerants require polyolester (ester), alkylbenzene, and polyalkylene glycol (PAG) lubricants of various viscosities. Additives will be used for certain refrigerants, such as HFCs.

Oil is regarded as a contaminant in used refrigerants[1,2]. This is because oil is a collection medium for contaminants, such as acids and particulates. Moisture levels in ester and PAG oils often exceed those in the refrigerant. Extra care is required when handling ester and PAG oils, as they absorb high levels of moisture when exposed to the atmosphere. Organic contaminants such as sludge, wax, and tars, can result from processing residues or mechanical wear when oil breaks down[3].

Acids in refrigeration systems may be organic or inorganic. Organic acids, frequently dissolved in the lubricant, can be formed by the breakdown of refrigerant or lubricant in the presence of moisture. More often, organic and inorganic acids are formed at high operating temperature conditions or when non-condensable gas levels are high. Generally, acids harm refrigeration systems[3]; the inorganics are considered the most harmful.

Metallic contaminants and dirt may be left in refrigeration systems during manufacturing, installation, or service[3]. Other particles can develop from wear or chemical breakdown during system operation. Particulates create problems with valve seats, plug orifices, and generally cause scoring and wear in compressor components (pistons, bearings, etc.).

# Removing Oil in Recovery-Recycling Units

Oil removal, which is very effective in that the oil contains acids and particulates, is the most important contaminant-removal step when recycling used refrigerant[4]. Separating oil facilitates moisture removal by use of filter-driers.

Kauffman[5] reported that oil separation, prior to passage of the used refrigerants through alumina-molecular sieve filter-drier systems, greatly improves the acid- and water-removal efficiencies of the recycling apparatus. When used with an inlet screen, the separator protects recovery system components such as solenoid valves, check valves, and expansion devices. Separating oil also protects the compressor oil sump from contamination and helps maintain the proper oil level. Finally, proper oil separation keeps contaminated oil out of the storage container where it is difficult to remove, often contributing to accelerated tank corrosion.

**Oil separation is essential when used refrigerant is to be introduced to systems without reclaiming. It may take longer to pump out vapor and separate oil, but clean recovery units, clean storage containers, and cleaner refrigeration systems are usually worth the extra time.**

## Simple Canister Separator

A canister may be used as a simple oil separator, Figure 4-1. When located after the evaporator, refrigerant vapor with entrained oil enters the separator through relatively small tubing. Upon entering the large-diameter canister, the refrigerant velocity is reduced. The velocity-reduction ratio is calculated by squaring the ratio of the smaller to the larger diameter. The compressor draws refrigerant vapor off the top of the canister while the heavier oil remains behind.

Design care is required to isolate the inlet from the outlet; otherwise velocity reduction may not be accomplished if refrigerant and oil take a direct path. A spiral path, such as around the canister wall, also may permit liquid oil to exit the canister outlet. However, when properly designed, the simple canister separator removes 90% of the oil.

## Heat Exchanger Separator

The evaporator and oil separator can be combined into a single unit, as studied in the previous chapter and shown in Figure 3-3. A distillation chamber[6] may be used to boil off refrigerant and separate oil, Figures 4-2 and 4-3.

Oils, Acids, and Particulates

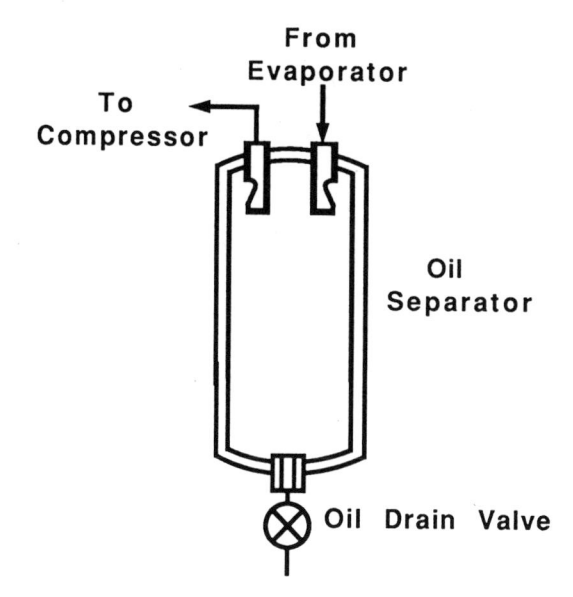

Figure 4-1. Simple canister oil separator

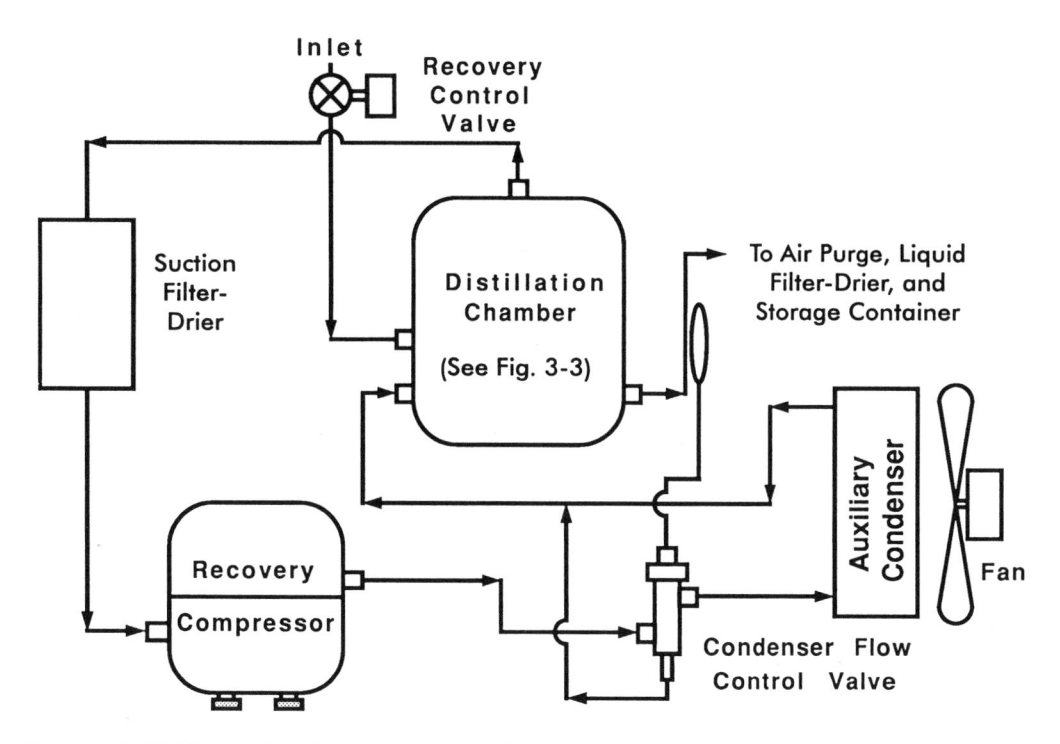

Figure 4-2. Distillation chamber oil separator schematic

47

# The Challenge of Recycling Refrigerants

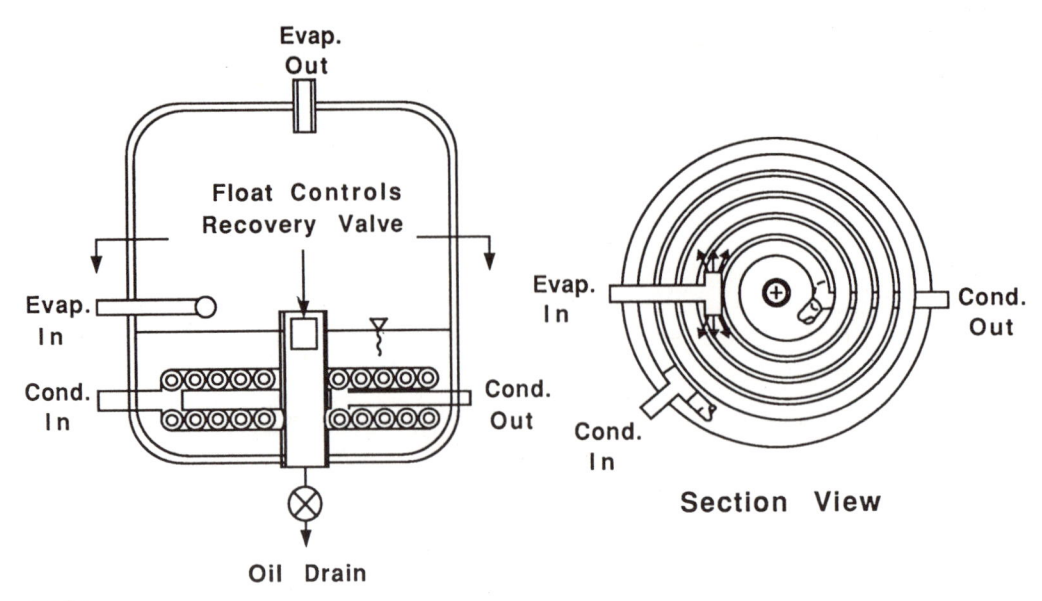

*Figure 4-3. Distillation chamber*

Figure 4-2 shows a chamber with a diameter of approximately 13 in. which separates more than 99% of the oil. The following special precautions are used to avoid turbulence and upward spiral oil migration (see Figure 4-3):

- Low-side float keeps the internal condensing coils covered with liquid refrigerant and lubricant.
- Recovery compressor pumps refrigerant out of the distillation chamber, which serves as an open evaporator.
- Primary condenser, located within the distillation canister, provides heat for boiling the refrigerant.
- Auxiliary condenser and flow control valve arrangement[7] provide maximum heat for boiling at low ambient temperatures while limiting discharge pressure at high ambient temperatures.
- Condenser flow control valve, using a sealed bulb at the chamber condenser outlet, provides no flow through the auxiliary below 100°F (38°C), partial flow between 100°F (38°C) and 120°F (49°C), and full flow above 120°F (49°C).

This single-pass recycling system results in refrigerant of reclaim quality[1] when processing R-22, using the ARI 740 contaminated sample and test method.

## Commercial Oil Separators

Commercially available oil separators using coalescing elements or spiral flow designs, can consistently remove oil to levels below 0.01% of the refrigerant weight[1] in a single pass. They are applied in the same way as the simple canister studied earlier.

## Draining Oil

In order for the separator to not be overfilled, it is important to determine when to drain off oil. For stationary equipment, recovery-recycling units must work on both smaller and larger systems. A sight glass in the canister can provide visual observation, or a liquid level sensor can be used. The compressor must pump down the oil separator while the recovery control valve is closed to avoid draining excessive refrigerant with the oil. This is especially important for distillation units. Oil separators should be drained after each recovery use and as required during long recovery operations.

## Oiling the Compressor

The recovery compressor oil reservoir must be maintained at the proper level and be contaminant free. Two oil separators are used in vapor pumpout units as shown in Figure 4-4. The inlet screen keeps the valve seats and orifices clean. The system oil separator, located between the evaporator and compressor, is important for contaminant removal and will need to be drained frequently. The compressor return oil separator serves two functions: 1) it collects any oil from the compressor discharge that would otherwise transfer to the storage container; and 2) it returns the collected oil to the compressor oil sump. (Oil return systems and compressor unloading were discussed in Chapter 3.)

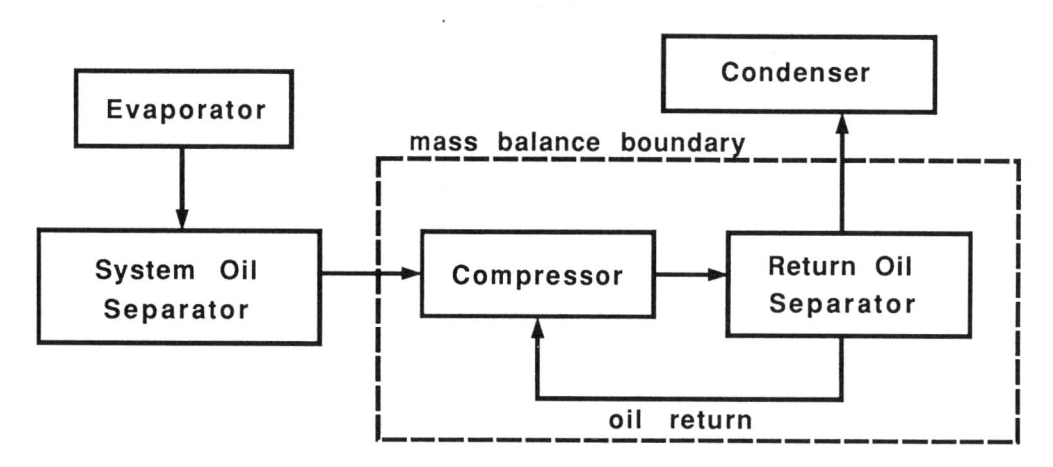

*Figure 4-4. Dual oil separators*

The dashed boundary line in Figure 4-4 shows that a mass oil balance on the compressor and return oil separator can be performed. Very little oil enters the compressor suction due to the system oil separator's high efficiency. Also, very little oil leaves the return oil separator outlet; most of it migrates from the compressor to the return separator, where it must be returned to the compressor.

Oil-less compressors do not require a return oil separator and oil return line. Good design practice still calls for balancing the compressor load before start-up.

## Liquid Transfer Considerations

Practical liquid-phase oil separators are not available. Therefore, oil separation cannot be accomplished when using the storage container cooling, push-pull, or liquid pumping methods described in Chapter 3. In such units, there is a temptation to omit the suction oil separator for the final vapor pumpout through the recovery compressor. This is poor economy. Once introduced into a storage container, oil is difficult to remove.

An apparatus to remove oil, shown in Figure 4-5, includes all the elements — flow control, evaporator, compressor, condenser, and oil separator — of a vapor pumpout unit. Yet, when the oil-free refrigerant enters the storage container, it immediately is diluted with oil-rich refrigerant. It is easier to keep oil out of a storage container than to subsequently remove it.

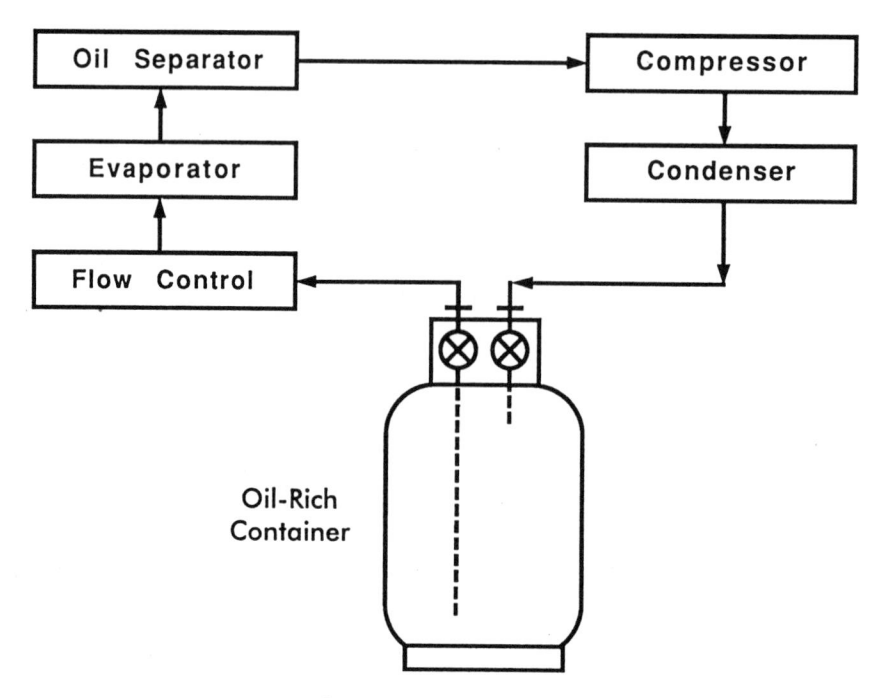

Figure 4-5. Recirculation to remove oil from container

Those wishing to remove refrigerant without separating oil should subsequently process the refrigerant through recycling equipment (as shown in Figure 4-2) and into a clean container. The recovered refrigerant-oil mix may also be sent to a reclaimer. In either case, the storage container must be cleaned.

# Tank Cleaning Procedure

The following procedure, recommended by Omega Recovery Services, Inc., for cleaning a storage container, is used with permission.

## General Cleaning

The normal used-refrigerant tank has a coating of oil and moisture on the inside walls of the tank. *All refrigerant should be removed from the tank prior to any cleaning operation.*

If there is not an acid condition, the simplest method is to pull a vacuum (at least 15 in. Hg) on the tank. Then attach a hose to the vapor tank valve.

Prepare a solution of either R-11 or R-113 mixed with isopropyl alcohol. Note that isopropyl alcohol is a flammable liquid and should be handled with proper care. This solution should contain not less than 25%, nor more than 40% alcohol. The alcohol is used to absorb and remove any moisture within the tank, as neither R-11 nor R-113 is adept at moisture removal.

The tank hose is then placed in a container (for 30- to 50-lb tanks, this could be a 5-gal bucket containing at least 2 gal of solution) holding the rinse solution. *Note: The hose should not have self-closing valves at the end connections. This will prevent the hose from passing the rinse solution into the tank.* Slowly open the tank valve to pull at least 1 gal of the solution into the tank. Shake or roll the tank so that all the sides are thoroughly washed with the rinse solution. Let it settle and then do it again. Afterwards, turn the tank upside down and drain the rinse solution from the vapor valve.

At first, the liquid coming out usually has an amber color. Properly dispose of the liquid rinse solution to comply with EPA and local regulations. Repeat as necessary until the liquid leaving the tank is reasonably clear. Then pull a vacuum on the tank to 25 in. Hg (635 mm). At this point, the inside of the tank is relatively clean of any oil and moisture contamination. *Caution: This procedure does not provide a tank that will maintain ARI 700 standards on newly reclaimed refrigerant but will be clean for storage of good, operating refrigerant, without additional contamination, and suitable for handling a variety of hvac applications.*

## Acidic Refrigerant Containers

If the tank stored any type of acidic refrigerant, the tank will require special cleaning procedures best addressed by proper technical companies. The longer the acidic refrigerant is stored in a steel tank, the greater the opportunity for the acidic condition to penetrate the initial metal surface of the tank.

Even after the acidic refrigerant is removed from the tank, any other refrigerant placed into the tank may absorb a certain amount of acid from the metal walls of the tank. The worse the acidic condition of the refrigerant, the higher the opportunity for the walls of the tank to absorb acid.

Attempting to determine if the refrigerant is acidic by using pH paper obtained from laboratory supply houses will not prove successful. Any reading by the pH paper test will provide an erroneous reading. It is best to have the refrigerant tested by a proper laboratory when an acidic condition is suspected.

Over time, used refrigerant tanks often hold a certain amount of "wet" refrigerant. When this happens, the surface area above the liquid level of the refrigerant begins to corrode. If the tank stores wide-ranging types and conditions of refrigerants, the valve should be removed and the tank visually inspected. All tanks must be reinspected at least every five years by an authorized DOT tank-inspection company. It is illegal to use or ship tanks containing refrigerants when the date code is more than five years old.

Aluminum valves, stems, or connections should not be used on steel tanks containing used refrigerant, as they accelerate any corrosion and contribute to an acidic condition in the tank.

# Disposing of Used Oil

Federal regulations on used oils presume that used oil containing more than 1,000 parts per million (ppm) halocarbons is a hazardous waste. Their concern is over certain solvent applications that truly are hazardous. If it can be shown by testing that halocarbons contain common refrigerants removed from air conditioning or refrigeration systems, the 1,000-ppm presumption is considered "rebutted" and the oil is not a hazardous waste.

Since the required test for each oil batch cannot practically be performed by service mechanics, the regulations do not consider the oil to be a hazardous waste while it is on the way to an oil recycling center, where the test can be performed. All used refrigerant oil should be taken to an oil recycling center, as it almost certainly will contain more than 1,000-ppm halocarbon refrigerant.

## Oil Analysis

Oil analysis on a compressor oil sample using acid test kits will be covered in Chapter 11. The liquid sample for analyzing refrigerants for high-boiling residue (oil), acids, chlorides, and particulates is the same as will be described in Chapter 5. A competent laboratory experienced in refrigerant analysis will analyze for:

- acids using titration;
- chlorides using turbidity;
- particulates using visual observation;
- high-boiling residue using volumetric methods.

# Future Design Considerations

It is anticipated that refrigeration system designers will incorporate positive-oiling systems with high-quality return oil separators in responding to different lubricant types, the use of zeotropic blends, and possible future retrofit considerations. Then, miscibility becomes less important, and it will become much easier to measure or change the oil as required. Alternatively, oil-less compressors may be used, so no lubricant is required.

# Summary

When recycling used refrigerant, oils, acids, and particulates can be collectively referred to as dirty oil. Removing oil is the most important contaminant-removal step when recycling used refrigerant. Oil separation is essential where used refrigerant is to be reintroduced to systems without reclaiming.

Oil separators are located in the recovery-recycling equipment suction line between the evaporator and the compressor, or they may be combined in function with the evaporators.

For proper oiling of the recovery compressor, an additional oil separator, located between the compressor and condenser, collects and returns oil to the compressor. Liquid transfer recovery units, such as push-pull models, cannot separate and remove oil.

Current tank cleaning procedures to remove dirty oil are not practical, as they involve the use of chemicals not readily available to the service technician. Used oil must be taken to an oil recycling center.

The amount of oil in a used-refrigerant sample can be determined by a competent laboratory.

# References

[1] ARI Standard 700-93, 1993, *For Specifications for Fluorocarbon Refrigerants*, Copyright 1993.

[2] ARI Standard 740-93, 1993, *For Performance of Refrigerant Recovery and/or Recycling Equipment*, Copyright 1993.

[3] ASHRAE, *1994 Handbook*, Chapter 6.

[4] *The Air Conditioning, Heating and Refrigeration News*, November 11, 1991 Issue.

[5] Robert E. Kauffman, ASHRAE Research Project 683-RP, 1993, *Sealed-Tube Tests of Refrigerants from Field Systems Before and After Recycling*.

[6] U.S. Patent 5,203,177, Robinair Division, SPX Corporation, 1993.

[7] U.S. Patent 5,261,249, Robinair Division, SPX Corporation, 1993.

# Chapter Five

# Moisture

Moisture is normally dissolved in the refrigerant or lubricant, but sometimes free water is present. Moisture[1] can come from exposure to the atmosphere, improper evacuation, improper manufacturing or installation practices, or from wet refrigerants and lubricants. Moisture also enters non-hermetic systems via permeation of non-metallic hoses and seals. Free water can come from leakage of water-cooled heat exchangers, or it can drop out of saturated refrigerants when the temperature is lowered.

Excess moisture[1] causes ice formation in orifices or valve seats; corrosion of metals and copper plating; and chemical damage to insulation in hermetic compressors or other system materials. The damage caused by excessive moisture increases with temperature or when air is present. Moisture also corrodes the metal inside a storage container.

New ester and PAG lubricants absorb high levels of moisture. These levels can approach 1,500 ppm or more, which is far more than the system chemistry can tolerate.

Moisture is removed from refrigeration systems by deep evacuation and by installing filter-driers. In some cases, dry nitrogen or dry air is piped through a wet system to remove excess moisture. Heat should be applied wherever possible to enhance the evacuation procedure or the dry air-nitrogen purge. In the past, "new" refrigerant was purged through the system to remove moisture, but this practice is no longer acceptable.

## Filter-Driers

Nearly all refrigeration systems have a drier to remove moisture from the circulating refrigerant. Small appliances use spun-copper driers. Automotive systems incorporate a bag of desiccant in the accumulator or receiver. Larger systems use one or more solid cores which can be easily replaced.

The Challenge of Recycling Refrigerants

Driers are normally placed in the liquid line between the receiver and the expansion device. Typically, a moisture-indicating sight glass is installed to determine the refrigerant's relative wetness. Suction line filter-driers are installed in particularly contaminated systems (as discussed in Chapter 11).

Filter-driers are typically of solid-core or loose-fill construction. The desiccant drying materials are usually molecular sieves or activated alumina. While there are distinctions within each type, the two basic materials will be studied. A solid-core filter-drier is shown in Figure 5-1.

*Figure 5-1. Solid-core filter-drier (Courtesy, Sporlan Valve Company)*

Cores are made of activated alumina, molecular sieves, and a binder material. The hollow core, which also serves as a particulate filter, is typically sealed at both ends. Refrigerant is introduced through the side wall of the metal shell, first flowing around the outside of the core, then through the core wall (where filtering and drying occur), and finally through an end opening.

A loose-fill drier is shown in Figure 5-2. The desiccant material is compacted under a spring load to contain the desiccant and minimize attrition. A screen and felt pad are typically used for filtration.

Moisture removal by desiccants tends to follow an equilibrium condition when the desiccant is exposed to continuously circulating refrigerant. As the desiccant adsorbs water, the drier reaches a maximum water capacity for a given endpoint refrigerant wetness and temperature. At higher endpoint wetness levels, the water capacity increases. Typical equilibrium curves for molecular sieves and activated alumina are shown in Figure 5-3.

Moisture

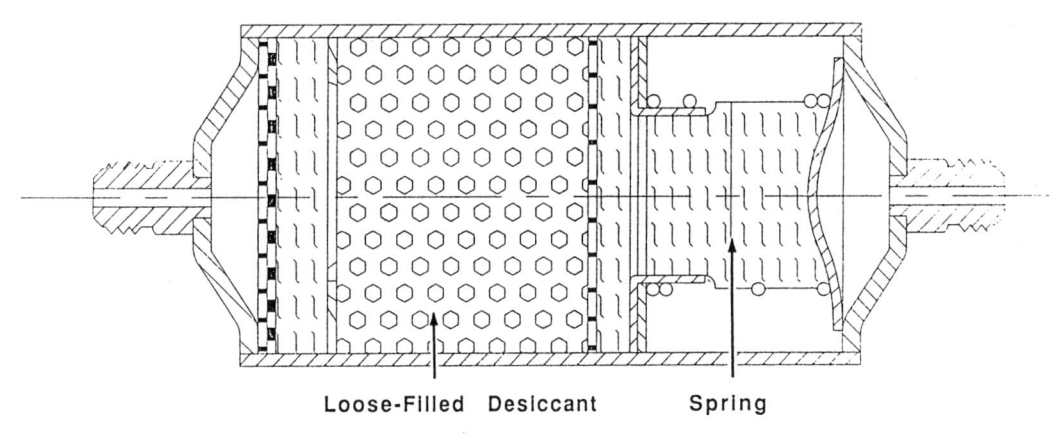

Figure 5-2. Loose-fill filter-drier (Courtesy, Sporlan Valve Company)

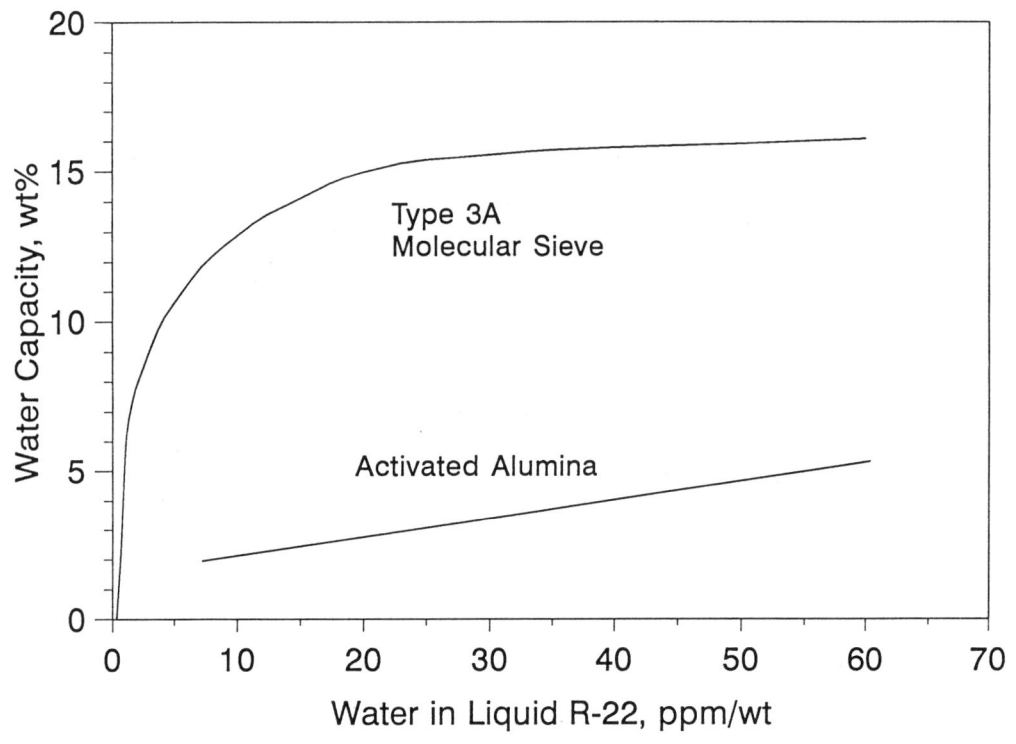

Figure 5-3. Moisture equilibrium curves for R-22 for two common desiccants (Courtesy, UOP)

57

Molecular sieve desiccants hold more than 15% of their weight in water when the endpoint wetness is greater than 20 ppm. The water-holding capacity is significantly less when drying below 10 ppm. For endpoint refrigerant wetness below 50 ppm, molecular sieves have considerably more water capacity than activated alumina.

Acid adsorption is not an equilibrium condition. There are currently no standards for rating acid removal capacity of driers. It is known that activated alumina has more capacity for removing certain organic acids.

## Recycling Equipment Filter-Driers

In recycling equipment, the primary function of the filter-drier is to remove moisture. Kauffman[2] found that oil separation, prior to passage of used refrigerant through the filter-drier system, greatly improved acid and water efficiency. Therefore, *the drier is best located after the system oil separator*. In this location, the recycling filter-drier does not contend with oil masking. Nor does it have to remove contaminants such as acid carried with the oil.

Multiple-pass filter-drying recycling units have been used successfully and have certain advantages for non-condensable removal. A schematic is shown in Figure 5-4.

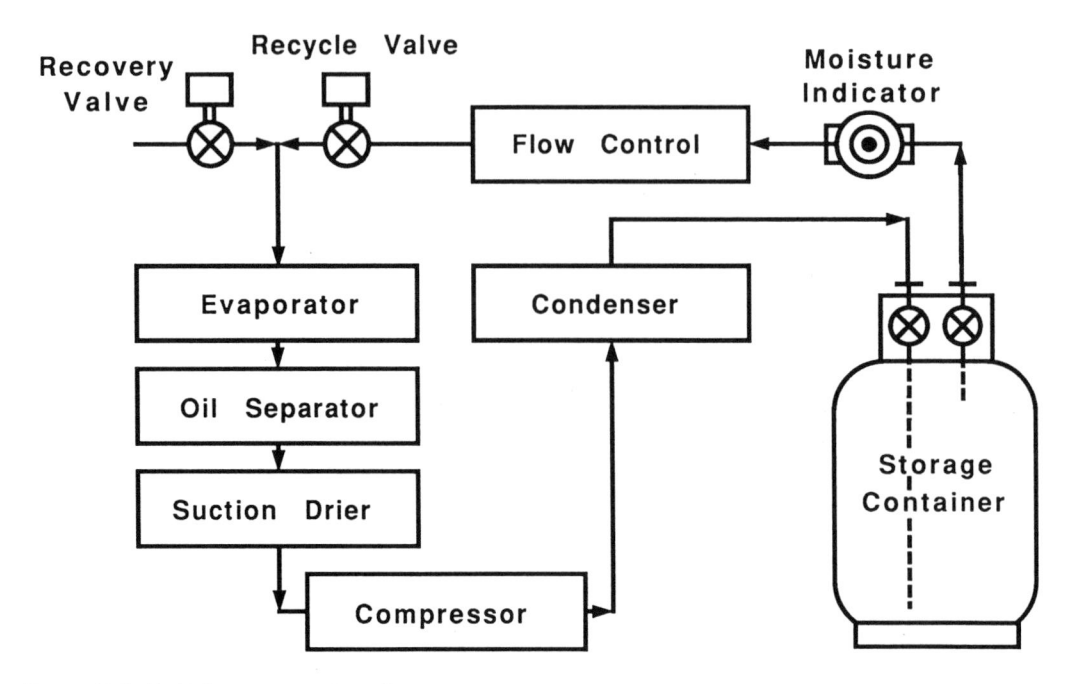

*Figure 5-4. Multiple-pass recycler schematic*

Single-pass filter-drying is currently preferred. Moisture removal by the drier in a single pass has less time to reach equilibrium endpoint wetness compared to refrigeration systems where multi-pass filter-drying is employed. In single-pass filter-driers, the inlet tends to be wetter than the outlet, and a wet-dry "front" travels from the inlet to the outlet as the drier capacity is used up. It is necessary to minimize dry pockets or corners to maximize water capacity.

Filter-driers suitable for refrigerant recycling unit usage are normally constructed of loose-fill molecular sieve beads or of solid-core construction with high molecular sieve content. Most filter-drier manufacturers have developed a "recycling" core for this application. Loose-fill driers normally have a longer flow path and higher capacity per gram of desiccant.

Loose-fill driers for a particular recycling unit may only be available through the recycling equipment manufacturer. Drier cartridges or replacement cores, available in the same geometry, may significantly differ in the way moisture is removed and in water capacity. Therefore, it is important to use the manufacturer's recommended type to achieve the proper results.

## Suction Filter-Driers

Suction filter-driers, shown in Figure 5-5, are used successfully in recycling units. The suction drier keeps the compressor and compressor oil sump dry and free of inorganic acids. When changing the drier, the compressor can be used to pump out refrigerant first.

While the flow velocity through a suction drier is much higher, the mass velocity contacting the desiccant is the same as for a liquid drier. To achieve an acceptable pressure drop, the length of the desiccant may need to be restricted to 10 diameters.

## Liquid Filter-Driers

Liquid driers, shown in Figure 5-6, are also used successfully in recycling machines. In this traditional location, flow velocity and pressure drop are lower for the same mass velocity than for suction line locations. Longer flow paths, such as 30 or more diameters[3], are possible.

# Change Indicators

Driers are capable of reducing moisture-contaminant levels well below new refrigerant specifications (generally 10 ppm by weight[4]). The real task is to know when the drier reaches its water capacity and can no longer dry to desired levels. Refrigeration system manufacturers are concerned that technicians will not change filter-driers if reliable indicators are not provided. **The drier is a key component of the recycling**

# The Challenge of Recycling Refrigerants

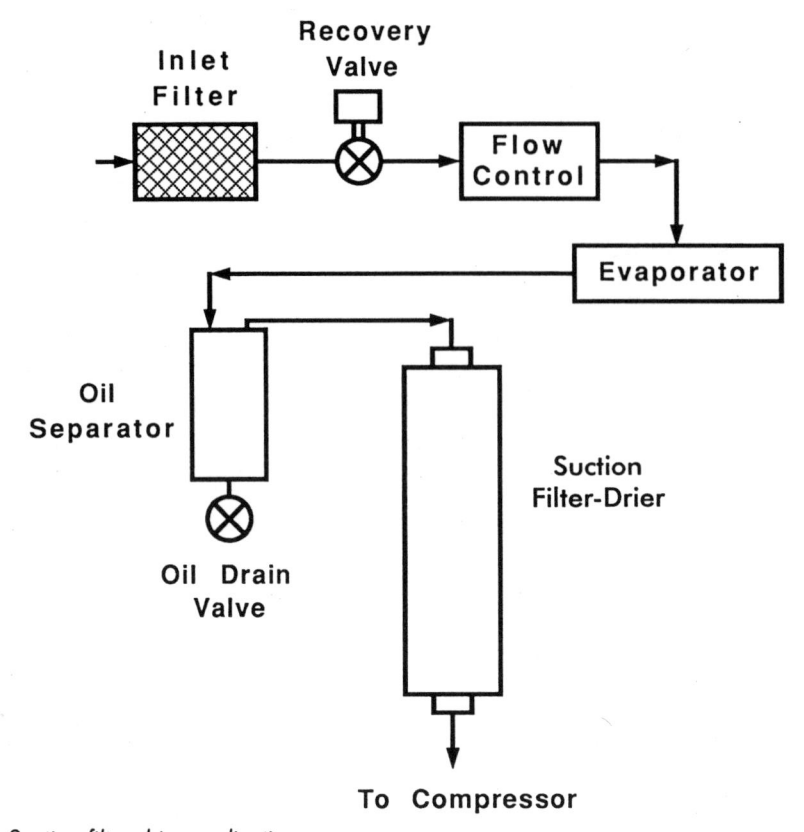

*Figure 5-5. Suction filter-drier application*

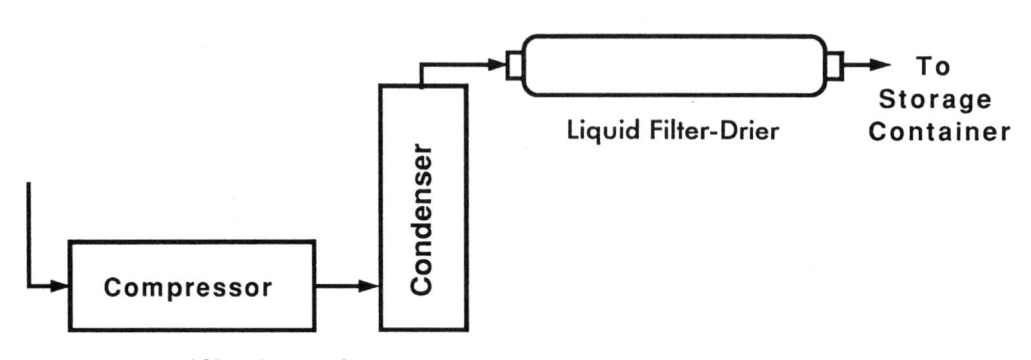

*Figure 5-6. Liquid filter-drier application*

**unit and must be changed as required or water will no longer be removed; moisture could even be added to dry refrigerant.**

Some have questioned how well recycling units will function after being used in the field for a long time. Recycling unit durability is similar to that of a refrigeration system. When the recycling unit runs and pumps refrigerant, the drier is replaced, and the oil separator drained, the unit is quickly restored to proper operating performance. When compared to new units, there is no reason to expect the performance to deteriorate if the unit is properly maintained.

## Chemical Salt Moisture Indicator

Moisture indicators, shown in Figure 5-7, have typically been used in refrigeration systems to measure refrigerant moisture levels. Located in the liquid line, they provide a color change to a "wet" condition at moisture levels well below free-water conditions, where icing could affect system operation.

In refrigeration systems, moisture levels change slowly (more week to week than minute to minute). The moisture indicator is only required to measure an average moisture level over several hours, usually with relatively clean refrigerant. These moisture indicators have been widely used in R-12 automotive recycling units. Moisture indicators show "wet" when R-12 exceeds 15 ppm, matching SAE[5] recycled refrigerant requirements. When the felt pad behind the salt dot is removed, the response time is much improved. (Normally, an indicator response time of less than 15 minutes is desired.) Except for occasional staining or washing of the salts, moisture indicators have worked reasonably well in R-12 applications.

Moisture indicators generally measure the relative wetness compared to saturated water levels for liquid refrigerant. Some refrigerants tolerate a lot more water than

*Figure 5-7. Chemical salt moisture indicator (Courtesy, Sporlan Valve Company)*

others before reaching saturation limits. For refrigerants such as R-22, R-502, and R-134a, the "wet" indication ranges from 40 to 80 ppm moisture by weight, which does not meet desired recycled refrigerant requirements.

Moisture indicators are not as reliable when processing the contaminated used refrigerants often encountered in stationary system applications. While chemical salt moisture indicators provide some protection against poor drying performance, they leave much to be desired as the primary drier-change indicator.

## Disposable Moisture Tubes

Glass tubes containing chemical salts may be used to determine refrigerant moisture levels as shown in Figure 5-8. The tube ends are broken and refrigerant flows through for a specified period of time. The length of the portion of tube changing color determines the moisture level. The tubes cannot be reused.

## Desired Moisture Transducer

There is currently a need for a real-time, in-line moisture transducer, which would measure moisture levels in the 10- to 20-ppm range[5] by weight. The transducer would be located where oil and other contaminants had already been removed. Either liquid- or vapor-flow locations are available. The response time would be less than 10 minutes, and the transducer would work for common refrigerants over a broad temperature range.

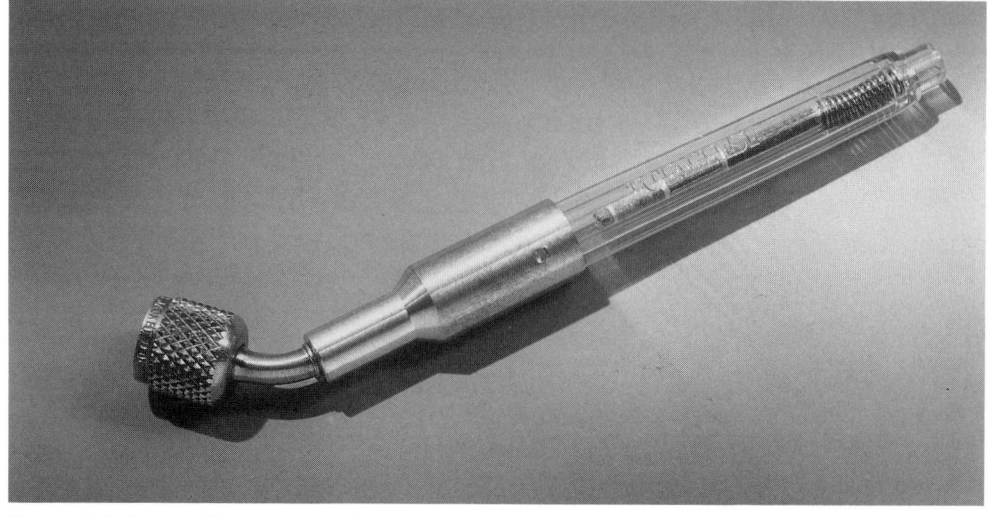

*Figure 5-8. Disposable moisture tube (Courtesy, Carrier Corporation)*

In operation, it could be "go" or "no-go," or a readout of actual moisture levels could be provided. As soon as the refrigerant at the drier outlet would exceed the setpoint, the recycling unit would indicate that the drier needs to be changed and, perhaps, stop the recycling operation.

While no such affordable transducer exists, companies and universities have studied the problem with varying degrees of success.

## Other Methods

Reliable filter-drier changeout indicators can be constructed based upon calculations involving saturated refrigerant moisture conditions and the drier moisture capacity. An hourmeter, volumetric flowmeter, in-situ mass flowmeter, and weight scale are a few of these methods.

When using a vapor pumpout recovery-recycling unit, a typical evaporator pressure profile is characterized by two phases, Figure 5-9. During phase 1, while liquid refrigerant feeds the recycling unit inlet, the evaporator temperature and pressure are at design conditions for the refrigerant type. During phase 2, the vapor refrigerant is removed until the final recovery vacuum is attained.

Both liquid and vapor densities significantly change with temperature for each refrigerant type. For vapor pumpout recovery units, suction line vapor conditions control flow rates. Refrigerant vapor density profiles are shown in Figure 5-10 for four common refrigerants using an evaporator temperature of 35°F (2°C).

The shape of the density and pressure curves in Figures 5-9 and 5-10 are similar, but notice the significant difference in density between the refrigerant types. Vapor moisture saturation curves[1] for three common refrigerants are shown in Figure 5-11. The oil separator, when used as shown in Figure 5-12, serves to limit moisture in the suction line to vapor saturation levels for the refrigerant.

### Hourmeter
The hourmeter measures recovery compressor run time. Calculations with an hourmeter must assume phase 1 flow (Figure 5-9) throughout. Referring to Figures 5-10 and 5-11, hourmeter calculations must assume the refrigerant type having the highest density and moisture saturation characteristics. Finally, the calculations must assume the highest suction pressure that design conditions allow. Expect to utilize 25% to 50% of the drier's water capacity when using an hourmeter.

### Volumetric Flowmeter
The volumetric flowmeter[6], such as that shown in Figure 5-13, can be used with a microprocessor to monitor refrigerant mass flow. Volumetric flowmeter calculations will be improved for phase 1 and phase 2 flow (see Figure 5-9) compared to an hourmeter.

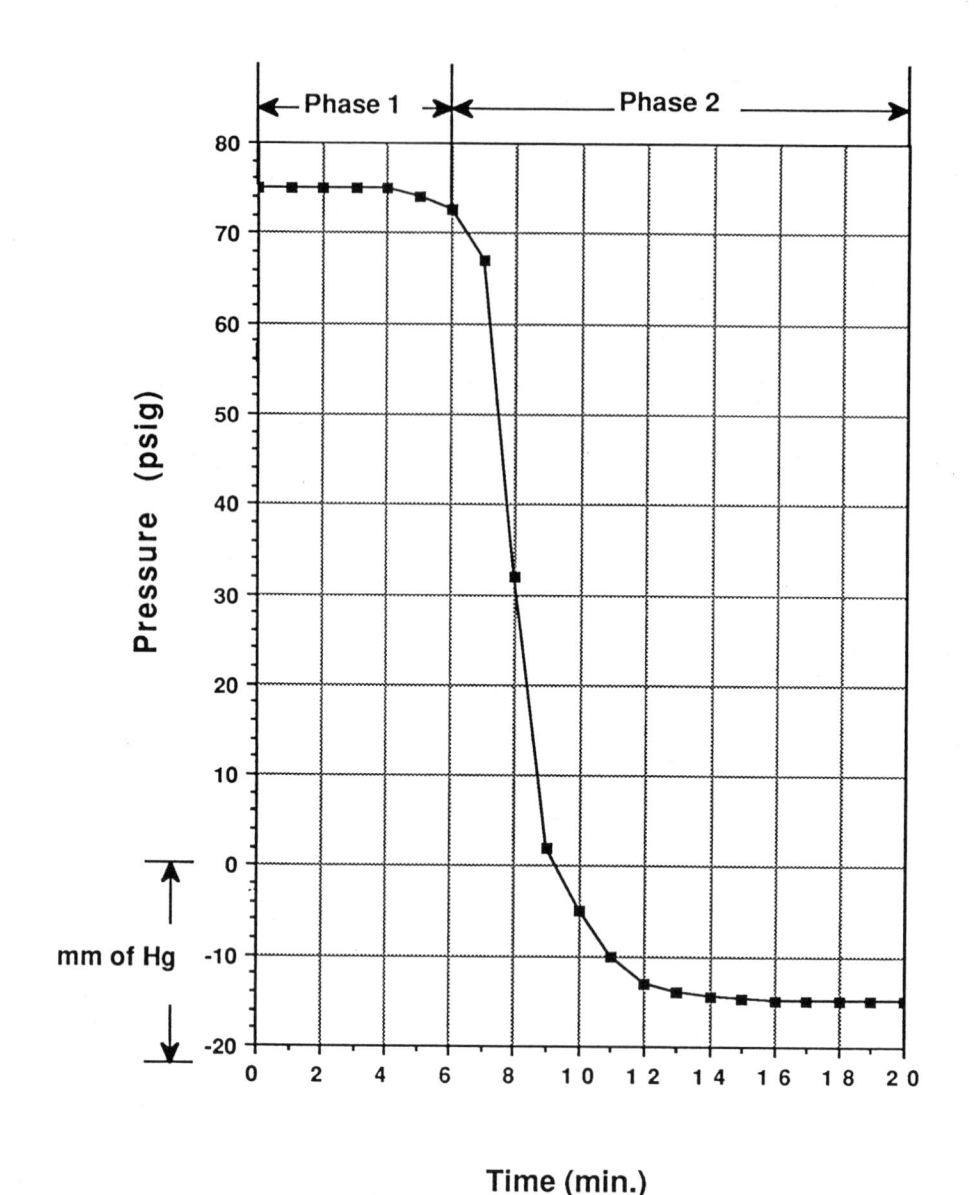

Figure 5-9. Typical evaporator pressure profile during recovery

The microprocessor calculates water adsorption based on moisture flow rate, which is calculated by multiplying volumetric flow rate by the "product" of density (Figure 5-10) and by moisture saturation (Figure 5-11). Since a less dense refrigerant like R-134a also has the higher moisture saturation (ppm) compared to a denser refrigerant like R-12, less moisture capacity is lost compared to hourmeter calculations. Expect to utilize 40% to 70% of the drier capacity using a flowmeter.

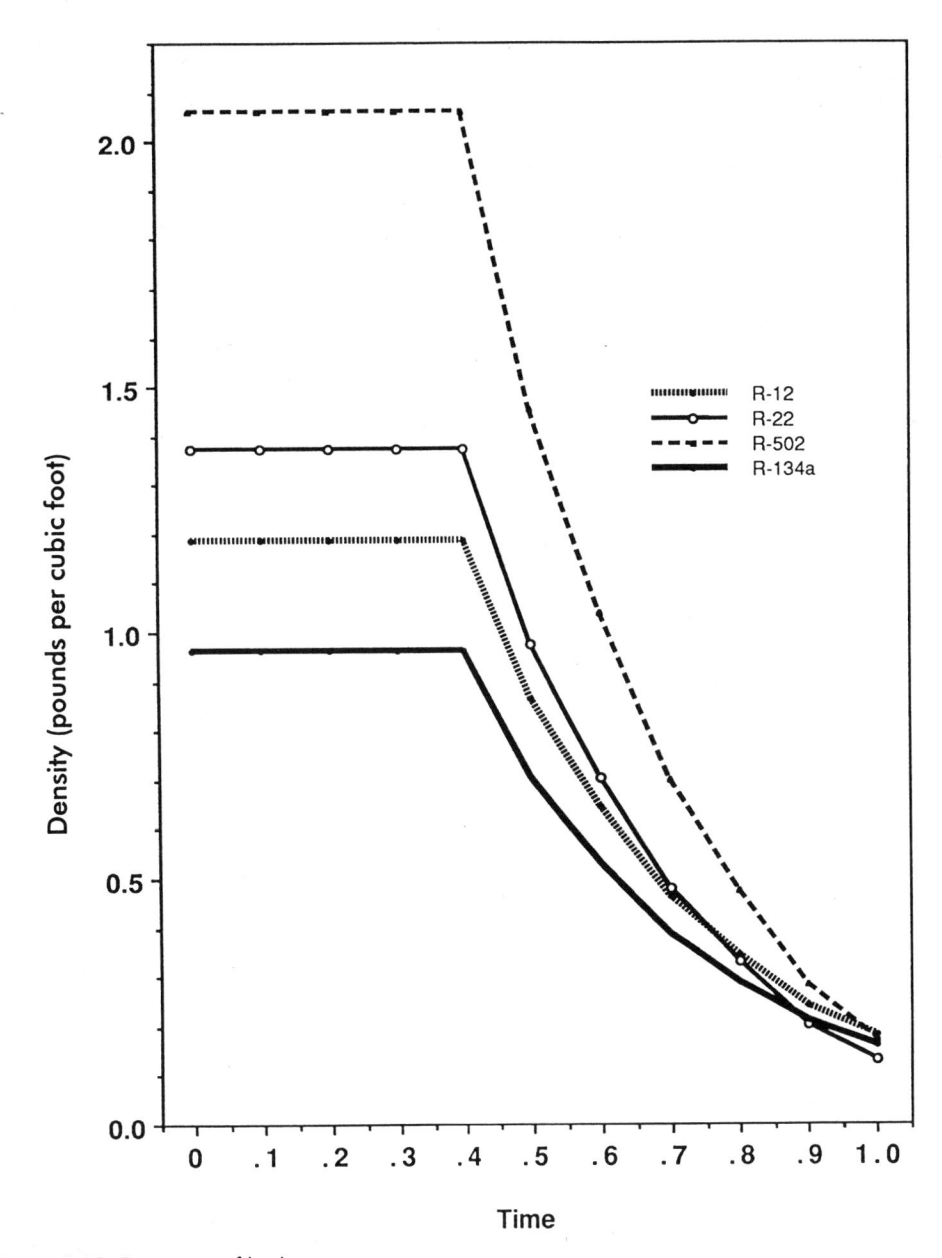

*Figure 5-10. Density profile during recovery*

### In-Situ Mass Flowmeter

The IS mass flowmeter[7] schematic is shown in Figure 5-14. The compressor, operating at constant speed, is used as a volumetric flowmeter. A temperature probe is placed at the evaporator, and a pressure transducer is placed in the suction line.

During phase 1 flow, the microprocessor uses pressure-temperature readings to determine the refrigerant type using look-up curves as shown in Figure 5-15.

## Moisture Solubility vs Pressure

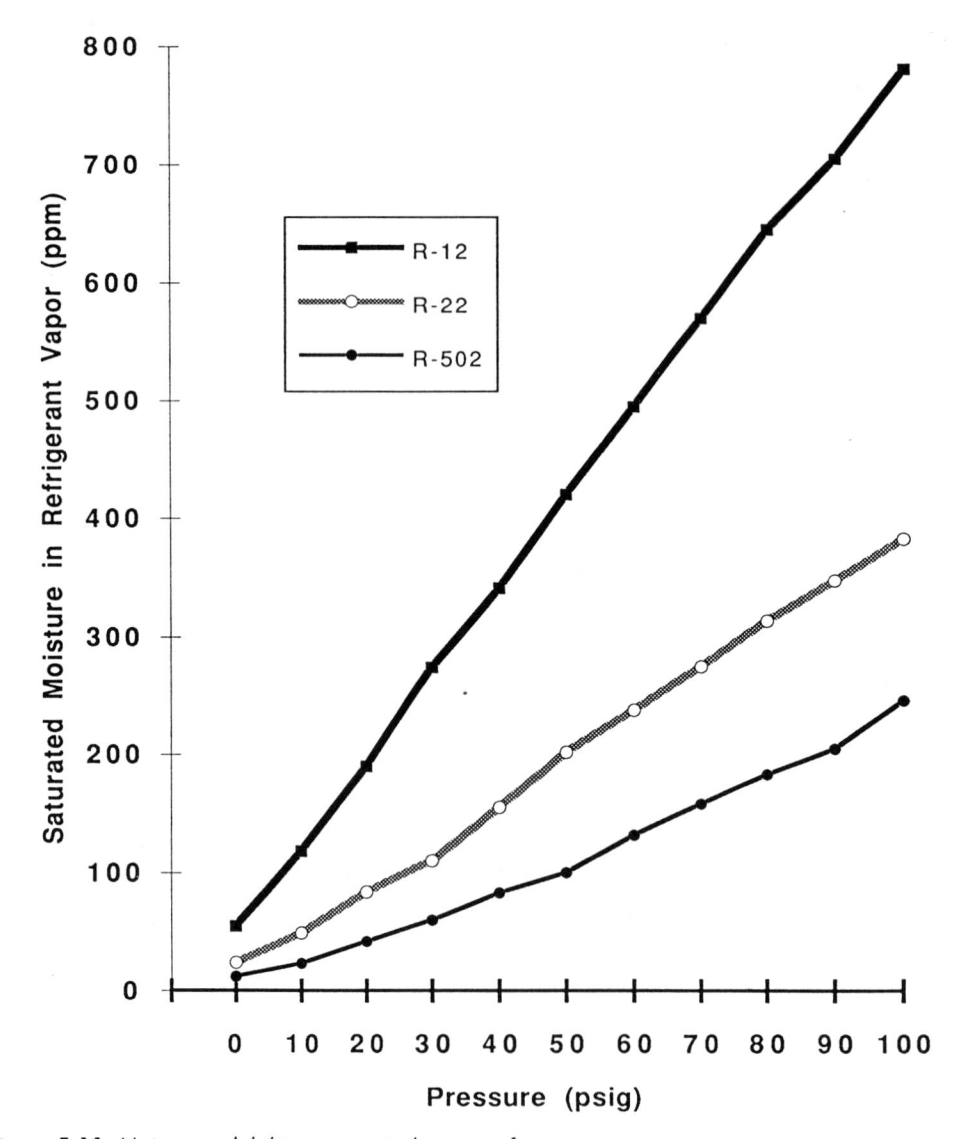

Figure 5-11. Moisture solubility in saturated vapor refrigerant

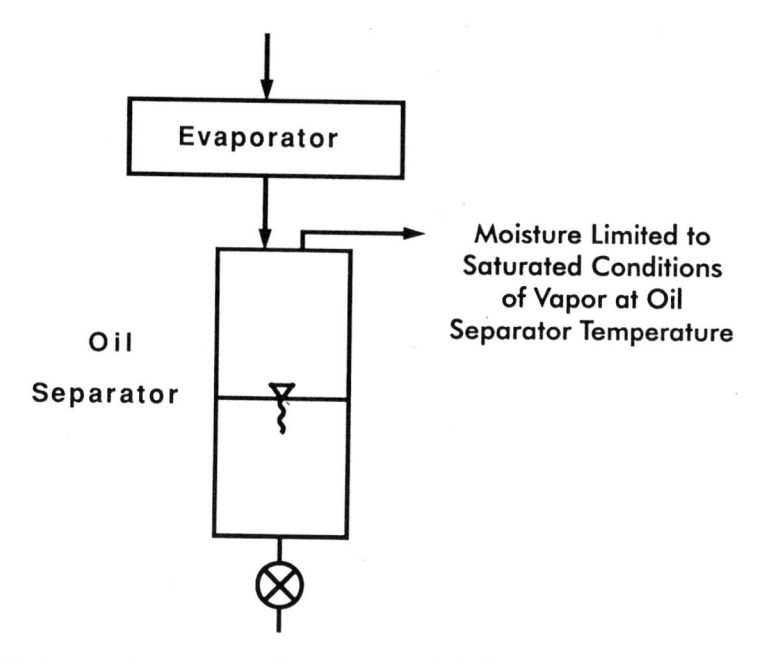

*Figure 5-12. Saturated moisture conditions produced at oil separator*

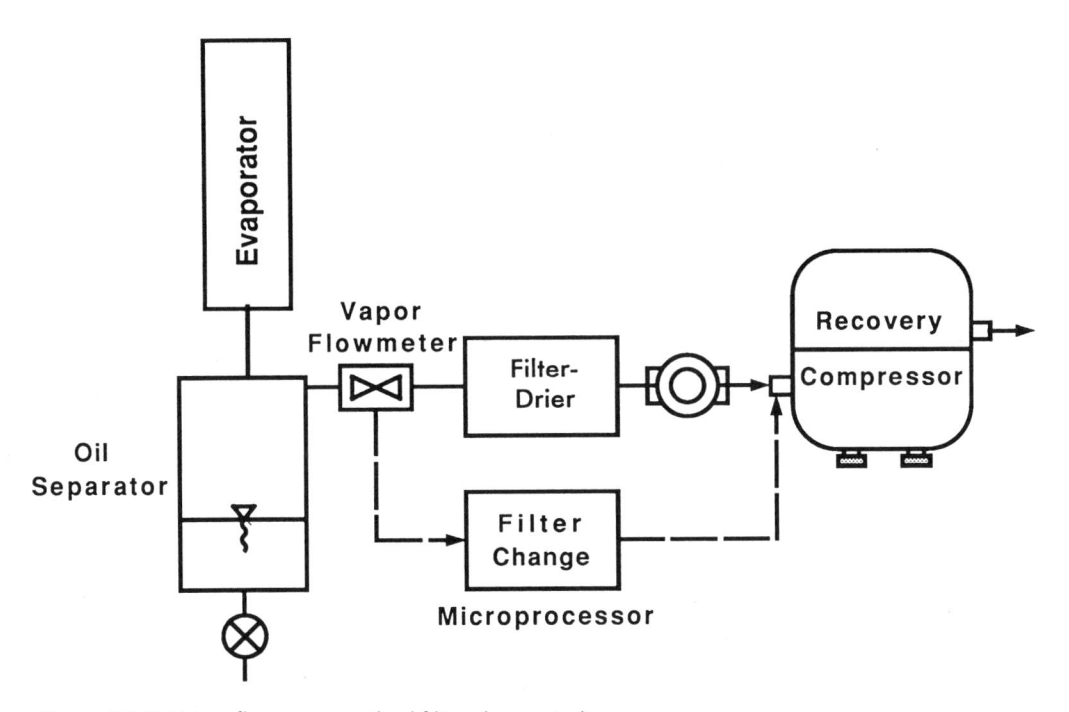

*Figure 5-13. Vapor flowmeter method filter-change indicator*

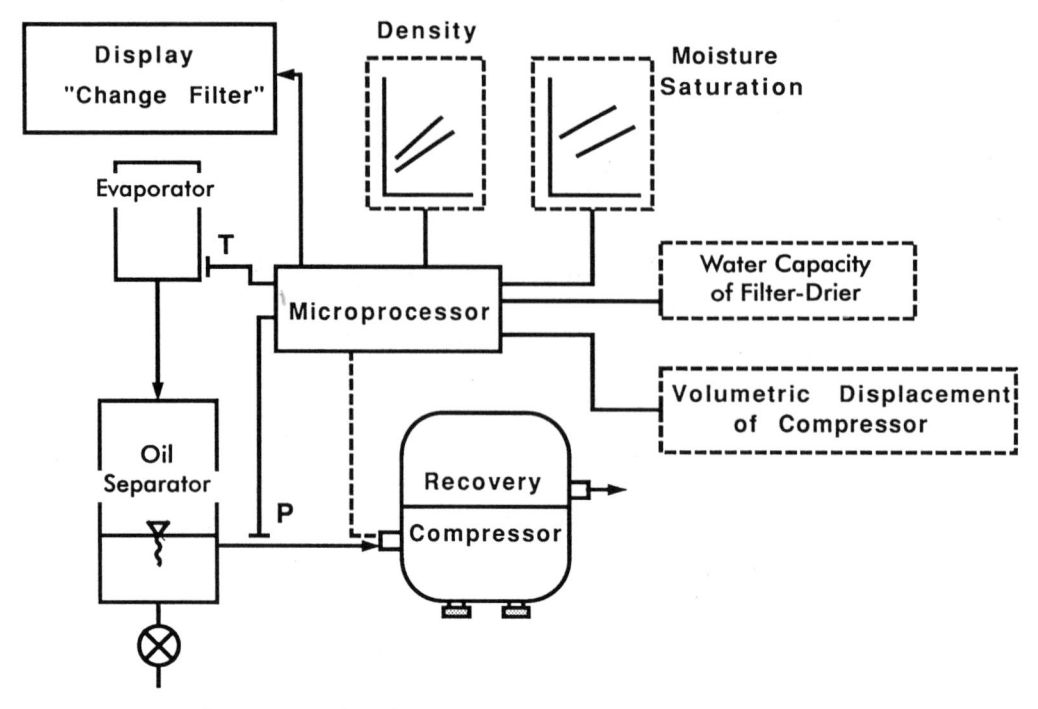

Figure 5-14. Mass flowmeter method filter-change indicator

Calculations using pressure (see Figures 5-9 and 5-10) provide accurate measurements of refrigerant mass flow for both phases 1 and 2.

Accurate saturated moisture calculations can also be determined (Figure 5-11) when the pressure is known. The oil separator ensures that saturated moisture conditions are not exceeded. The moisture flow rate calculations follow:

$$m_{H_2O} = (Ps)(\rho)(M_{sat})$$

where: $m_{H_2O}$ = Instantaneous moisture flow rate (lb/min)

Ps = Volumetric compressor pumping speed (cu ft/min)

$M_{sat}$ = Moisture saturation ratio (lb water/lb refrigerant)

$\rho$ = Refrigerant density (lb/cu ft)

## Evaporating Temperature vs Vapor Pressure

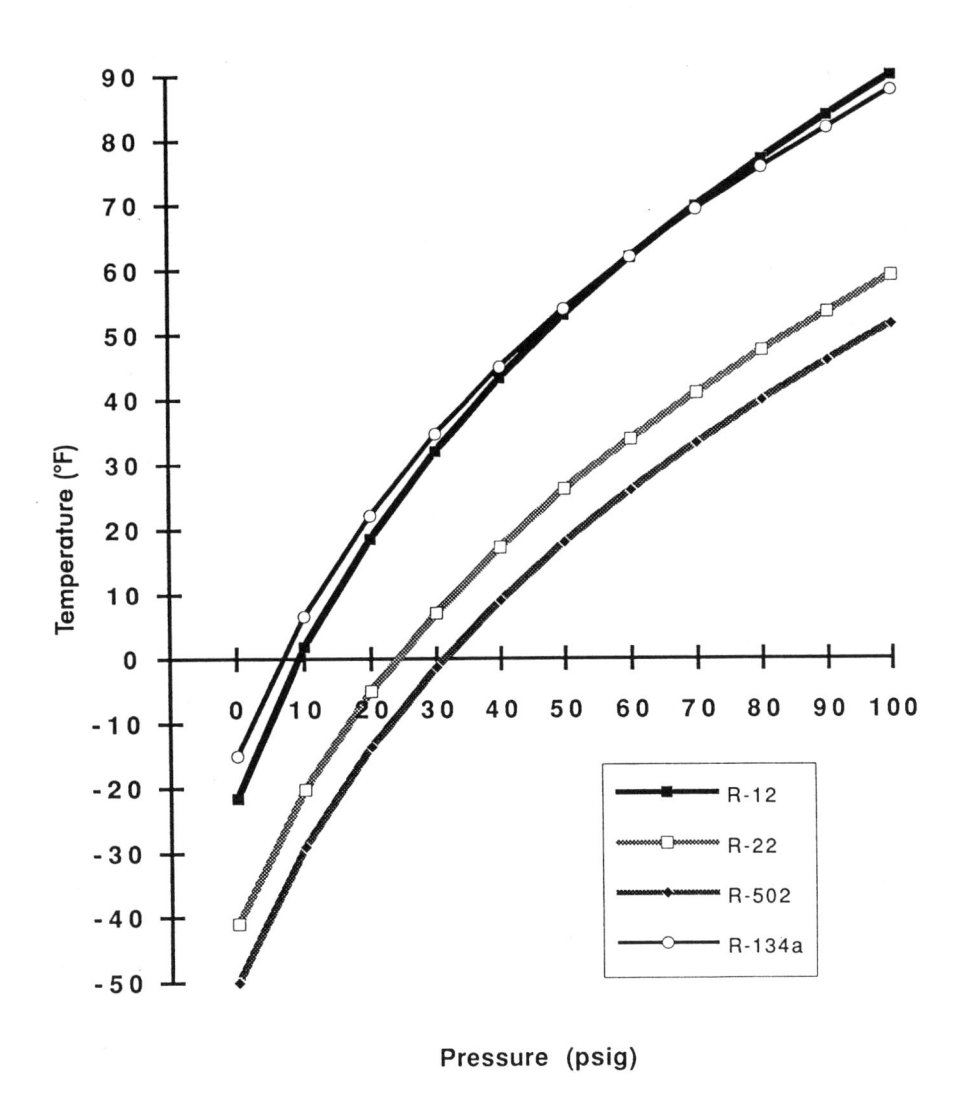

*Figure 5-15. Saturated pressure-temperature curves*

$M_{H_2O} = (\Sigma M_{H_2O})(\Delta t)$

where: $M_{H_2O}$ = Accumulative moisture removed (lb)

$\Delta t$ = Time interval over which $M_{H_2O}$ is measured

In other words, the water removed by the drier is the sum of all the small amounts of water removed in each small time interval. For saturated moisture conditions, expect to utilize 90% of the drier capacity. The expected range, using the in-situ mass flowmeter, is 50% to 90%.

**Weight Scale**
Some recovery units have a scale for weighing recovered refrigerants. Refrigerant trapped in the unit condenser cannot be weighed. For recycling units that use multiple-pass filter-drying, care must be exercised to weigh the same refrigerant only once. Removed water calculations are based on saturated moisture curves (Figure 5-11) based on the "wettest" refrigerant curve, unless the refrigerant type is entered or separately determined.

# Changing Filter-Driers

When changing filter-driers, care must be exercised to avoid venting refrigerant from the recycling unit and to remove air that has entered the unit while opened. The recycling unit can be used to pump refrigerant out of the drier before the changeout.

To change a suction drier, first close the recovery control valve (inlet) and pump down with the compressor. After changing the drier, use a vacuum pump to remove any air introduced. The compressor may be used (Figure 3-9) instead of a vacuum pump.

To change a liquid drier, valving and connections are required, Figure 5-16. First, close the isolation valves and connect a hose from the service port to the recycling unit inlet. Then, loosen the hose nut near the drier to purge air before starting the recovery unit to remove refrigerant. After changing the drier, pull a vacuum through the service port. Alternatively, the entire unit can be pumped down to remove the refrigerant as discussed in Chapter 3.

After careful review of federal regulations and consulting with chemists and filter-drier manufacturers, it has been determined that used driers, when used in recycling unit locations after the oil separator, can be discarded. Federal regulations are subject to change, however, and different state and local regulations may apply.

# Removing Free Water

Situations such as burst water tubes producing free water in chiller refrigerant circuits require special analysis and procedures to efficiently remove the water. For example, a water tube ruptured on-site during a recycling operation on a large chiller in Texas. The refrigerant had been pumped into a large receiver, then pulled out of the bottom of the receiver, recycled, and returned to the receiver.

When repeated refrigerant sample analyses showed the moisture level was not decreasing, it was determined that there was free water in the receiver. Another tank was brought in, and the recycled refrigerant was pumped into the extra tank. When

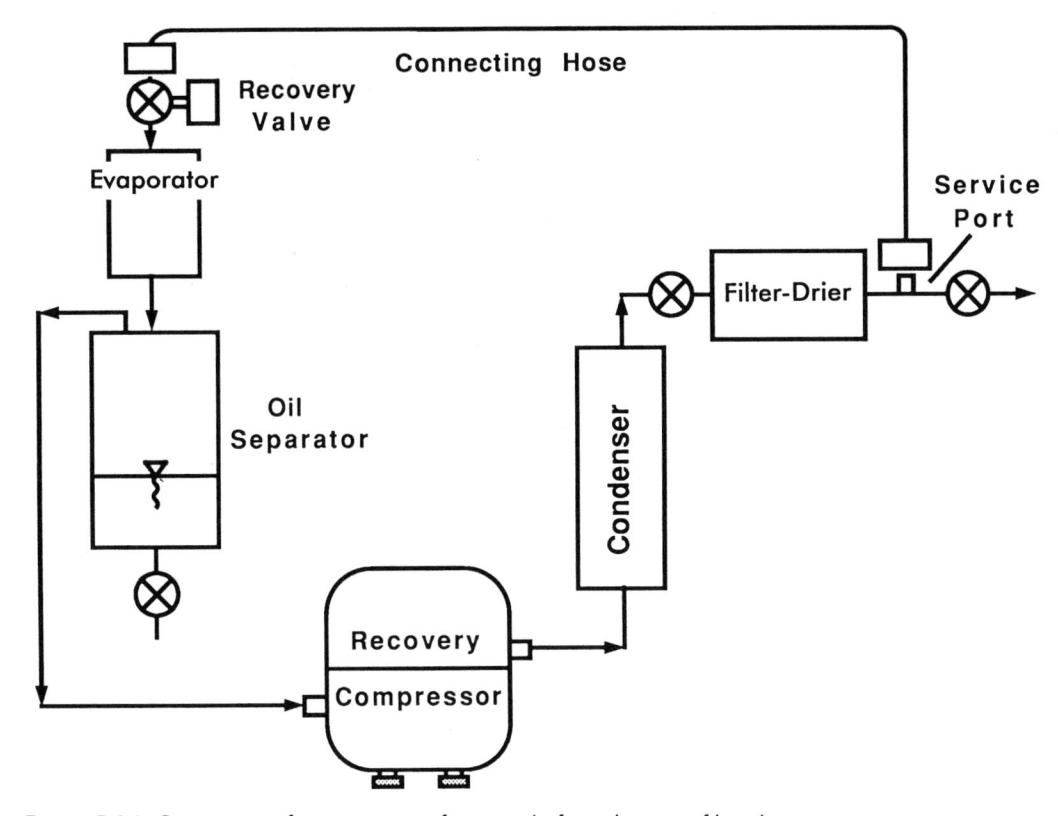

*Figure 5-16. Connections for removing refrigerant before changing filter-drier*

the receiver contained mostly water, the refrigerant was recovered from the vapor port, the water was drained, and the receiver dried out. In this manner, considerable time was saved.

## Laboratory Sampling

When sampling refrigerant for moisture or other contaminant analysis, consider the following points:

- Work with a competent laboratory that has experience in analyzing refrigerants.
- Exercise care to ensure the moisture content in the refrigerant sample is representative of the system or container.
- Ensure the tubing or hoses used to collect the sample and the sample container are clean and dry. Draw refrigerant sample from the liquid phase.
- Use the sampling procedure recommended by the laboratory (the laboratory should furnish or recommend the sample container and safe shipping instructions).

The laboratory will analyze refrigerant for moisture levels using a Karl Fischer Coulometer.

## Summary

Moisture causes serious problems in refrigeration and air conditioning systems. New ester and PAG lubricants absorb high levels of moisture when exposed to atmospheric conditions.

Filter-driers are used to remove moisture from refrigerants. Recycling applications are unique from system applications both in the filter-drier construction and its location in the equipment. Refrigeration system manufacturers are concerned that technicians will not change filter-driers if reliable indicators are not provided.

While the desired moisture transducer is not yet commercially available, reliable filter-change indicators can be used, although the full moisture-holding capacity of the filter-drier may not be used. Refrigerant should be removed before, and air should be removed after the replacement filter is installed.

A liquid sample is required for analysis by a competent laboratory.

## References

[1] ASHRAE, *1994 Handbook*, Chapter 6.

[2] Robert E. Kauffman, ASHRAE Research Project 683-RP, 1993, *Sealed-Tube Tests of Refrigerants From Field Systems Before and After Recycling.*

[3] U.S. Patent 5,240,483, Shelvin Rosen, 1993.

[4] ARI Standard 700-93, 1993, *For Specifications for Fluorocarbon Refrigerants*, Copyright 1993.

[5] SAE J1990, 1989, *Extraction and Recycle Equipment for Mobile Automotive Air-Conditioning Systems.*

[6] U.S. Patent 5,211,024, Robinair Division, SPX Corporation, 1993.

[7] U.S. Patent Pending, Robinair Division, SPX Corporation.

# Chapter Six

# Air and Non-Condensables

Air is a serious contaminant in refrigeration systems. It can cause higher head pressures, higher compressor discharge temperatures, and lower cooling efficiencies. Tolerance for non-condensable gases[1] depends on the system. Higher temperatures speed up undesirable chemical reactions; air (oxygen) contributes to the reactions, too.

Air found in refrigeration systems can result from incomplete evacuation or leaks. Low-pressure refrigeration (R-11) systems that operate in a vacuum may draw air through seals and joints. Purge systems are required to remove the air, and some of the older systems also purge refrigerant in so doing. Newer purge units, however, are much more effective in separating out the refrigerant before purging air.

In extreme cases involving excessive moisture and high temperatures, system chemistry may form non-condensable gases, such as hydrogen chloride. This, in turn, attacks components in the refrigeration system and may cause the system to fail[1] in extreme cases.

## Evacuation

Proper evacuation is essential to remove both air and moisture from refrigeration systems. These two contaminants are especially damaging to systems. They cause corrosion, acid formation, and higher compressor loading. Using a high-quality vacuum pump, Figure 6-1, and an electronic vacuum gauge, Figure 6-2, ensures proper evacuation.

The vacuum pump should be capable of pulling to well below 100 micrometers of mercury (microns Hg, or 0.002 psia). Two-stage rotary vane vacuum pumps work quite well for stationary systems, while single-stage vacuum pumps are sufficient for automotive systems. Vacuum levels are commonly expressed in micrometers of mercury (microns Hg) or pounds per square inch absolute (psia).

*Figure 6-1. Two-stage rotary vacuum pump (Courtesy, Robinair Division, SPX Corporation)*

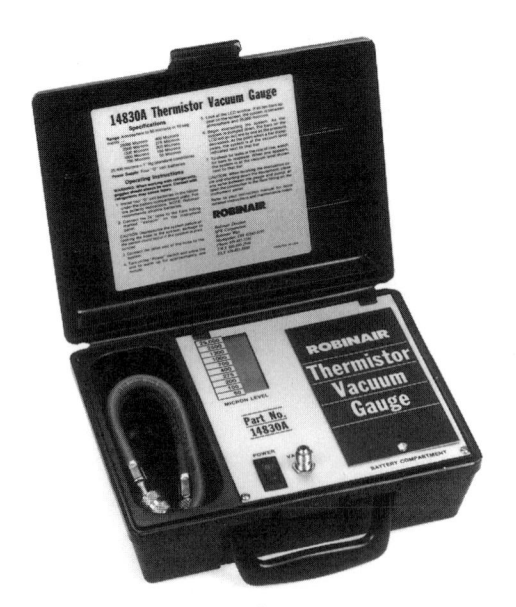

*Figure 6-2. Thermistor vacuum gauge (Courtesy, Robinair Division, SPX Corporation)*

*Vacuum pumps* are rated in free air displacement (cubic feet per minute or cubic meter per second) and final blank-off pressure (psia or microns). A 6-cfm (2.8-l/sec) pump can evacuate systems with up to a 50-ton (176-kW) refrigeration capacity.

*Rotary vane pumps* use an oiling principle to achieve deep vacuum. Initially, the maximum expanded volume fills with a mixture of rarefied gas and oil. As compression progresses, the pump reaches the point of solid oil fill, and then hydraulic expansion begins: The flapper valve opens early and gas molecules are carried out, entrained in the oil, causing extremely high compression ratios.

To remove moisture, pull a vacuum to 500 microns (0.01 psia) or less and hold for several minutes. Use the electronic vacuum gauge to ensure the pressure does not rise, indicating outgassing of refrigerants or leakage. An isolation valve on the vacuum pump allows checking for pressure rise in the system when the pump is isolated.

## Service Connections

The refrigerant-metering device (TXV) is a significant restriction to evacuating through only one port. Connecting to both high- and low-side service ports will shorten evacuation or refrigerant recovery time. Using short hoses of larger diameter further decreases the time required.

Connecting and then purging hoses has been a major source of frustration to service mechanics. Using a valved hose minimizes the air introduced into the system or service equipment, as well as the refrigerant lost to the atmosphere. The automatic valve hose shown in Figure 6-3 is especially convenient. A spring-loaded poppet opens the hose to flow when the hose nut contacts the service port face. When the hose is disconnected, the poppet closes and the hose seals.

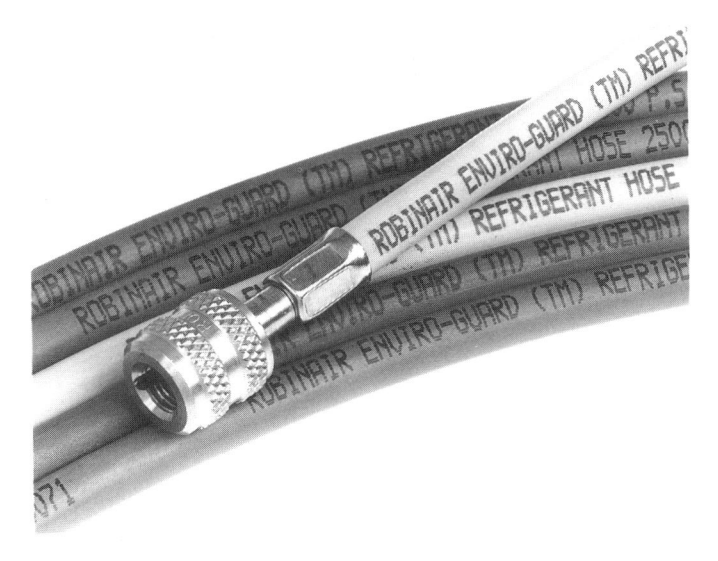

*Figure 6-3. Nylon barrier hose with automatic shutoff valve (Courtesy, Robinair Division, SPX Corporation)*

Low-permeation hoses (e.g., nylon barrier construction) minimize refrigerant leakage to the atmosphere. R-12 permeation rates for standard hoses (neoprene) are more than 40 times higher than for low-permeation hoses.

Hoses meeting SAE J2196 (motor vehicle air conditioners) or ARI 740-94 (stationary equipment) requirements should be selected. Special valves for automotive recycling units will be discussed in Chapter 10.

# Measuring Non-Condensables (NCs)

There are two different methods available for measuring NCs: the partial pressure method and the oxygen sensor method.

## Partial Pressure Method

Measuring partial pressure due to NCs is similar to measuring superheat. Pressure and temperature are measured in a location where both liquid and vapor refrigerant exist. The saturation pressure for the refrigerant is calculated from the temperature. The partial pressure due to NCs is the difference between the actual pressure and the saturation pressure.

New or reclaimed refrigerant may contain up to 1.5% NCs by volume. For R-12 at room temperature, the corresponding partial pressure of NCs is less than 1.5 psi (10.3 kPa). If the temperature were measured within 0.5°F (0.3°C) and the pressure within 0.5 psi (3.4 kPa), partial pressure could be determined within 1.5 psi (10.3 kPa).

In laboratory tests at Robinair, a storage container was outfitted with temperature probes, Figure 6-4. Probes T1, T2, and T3 were at the top, middle, and bottom, respectively, outside the tank. Probes T4 and T5 were at the top and bottom, respectively, inside the tank.

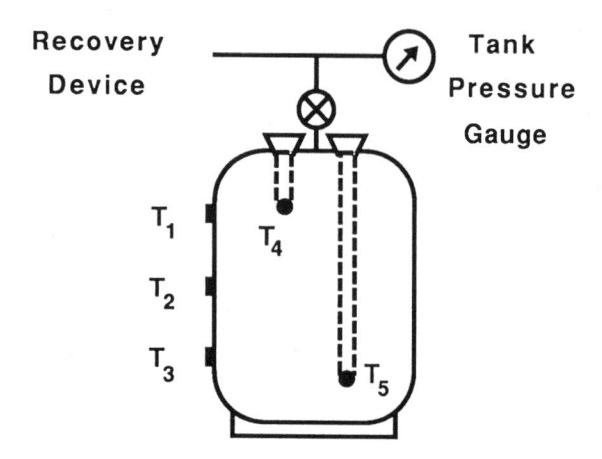

*Figure 6-4. Storage container with temperature locations*

As the tank filled during successive recovery operations, the temperature closest to the current tank-fill level was the most accurate. When the tank was 80% full, the bottom probe (T3) was 25°F (14°C) cooler than the top probe (T1). This shows the need to circulate the refrigerant: to achieve uniform tank temperature and accurately measure partial pressure of NCs.

### Oxygen Sensor Method

An oxygen sensor of the type used to measure oxygen levels in medical applications may be capable of measuring oxygen levels in storage containers. This direct measurement is unaffected by different refrigerants or pressure-temperature effects. In situations where nitrogen is used for leak testing, an oxygen sensor would not be effective. An oxygen (or nitrogen) sensor can be placed at or near the storage container. If the NCs exceed allowable levels, the purge solenoid valve opens to purge. This method is not sensitive to refrigerant type or temperature.

## Purging NCs in Recycling Units
### Manual Air Purge for Single Refrigerants

It is important to circulate refrigerant out of the storage container, past the air purge indicator, and back to the container to establish thermal equilibrium. The compressor or a separate liquid pump circulates the refrigerant. A manual valve purges vapor from the top of the container.

The air purge indicator consists of two pressure gauges. The first is connected to a sealed bulb containing the same refrigerant as the one being tested; the second measures the pressure in the storage container. The location of the sealed bulb must be optimized with each design.

The two gauges can be combined into a single unit, Figure 6-5. Two Bourdon tube gauges, mounted in a single housing using concentric shafts, use two indicator needles on a single dial face. Alternatively, a differential pressure gauge can be used with one port connected to a sealed bulb and another to the tank.

Preferably, the NC-measuring instrumentation will be placed on the recycling unit and not on the portable storage container. Then the storage container can be replaced without duplicating the instrumentation.

### Automatic Air Purge for Single Refrigerants

An automatic purge valve[2,3], applied as shown in Figure 6-6, is constructed by adapting a TXV. Refrigerant is circulated as previously described for temperature stabilization.

The Challenge of Recycling Refrigerants

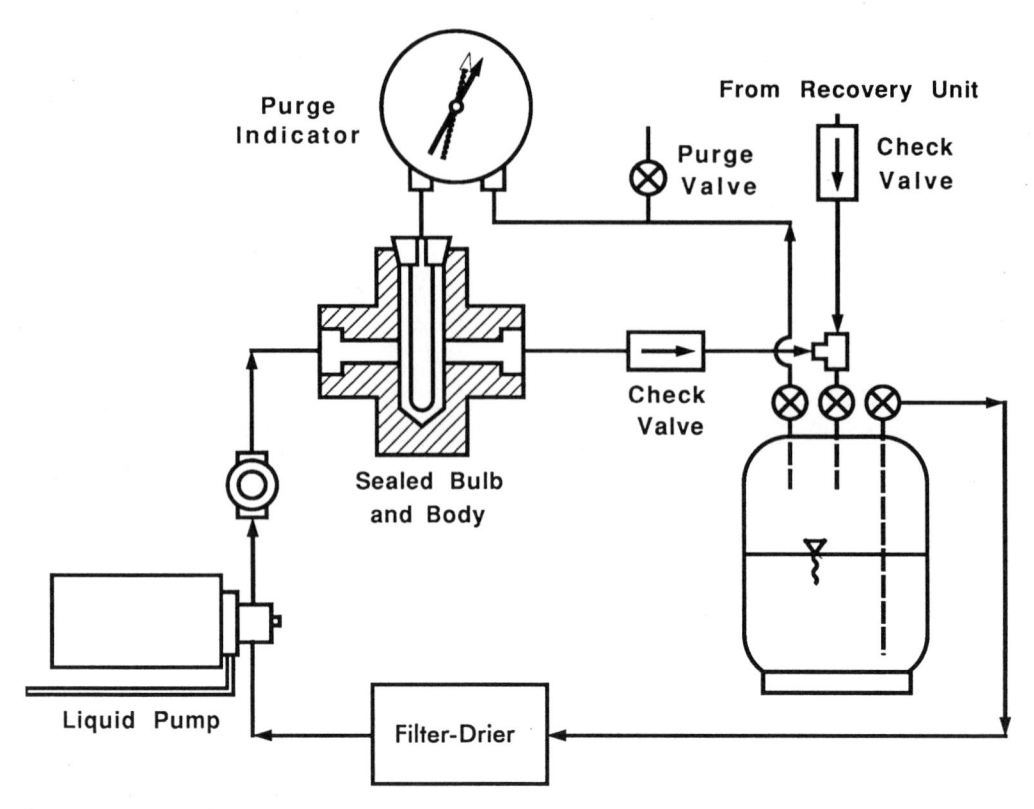

*Figure 6-5. Manual NC purge — single refrigerant*

Refrigerant contained in a sealed bulb located within the top "hat" section of the valve produces a downward or closing force on the valve. A spring, also in the hat, produces a closing force equal to the allowable NC partial pressure before starting to purge.

The tank pressure causes an upward or opening force on the valve. A solenoid valve prevents purging when temperatures have not stabilized. The valve automatically purges when the valve is open and the NC partial pressure exceeds allowable levels.

## Manual Purge for Multiple Refrigerants
The dual-needle pressure gauge[4], shown in Figure 6-7, is designed for R-12, R-22, and R-502. Like the gauge in Figure 6-5, one needle reads tank pressure, which is shown on the outer scale. The other needle reads R-22 bulb pressure. Pressure scales are provided for all three refrigerants.

The manual air-purge valve is opened until the tank pressure on the outer scale matches the refrigerant saturation pressure on the appropriate inner scale. This design is limited to three or four refrigerants, due to dial face layout constraints.

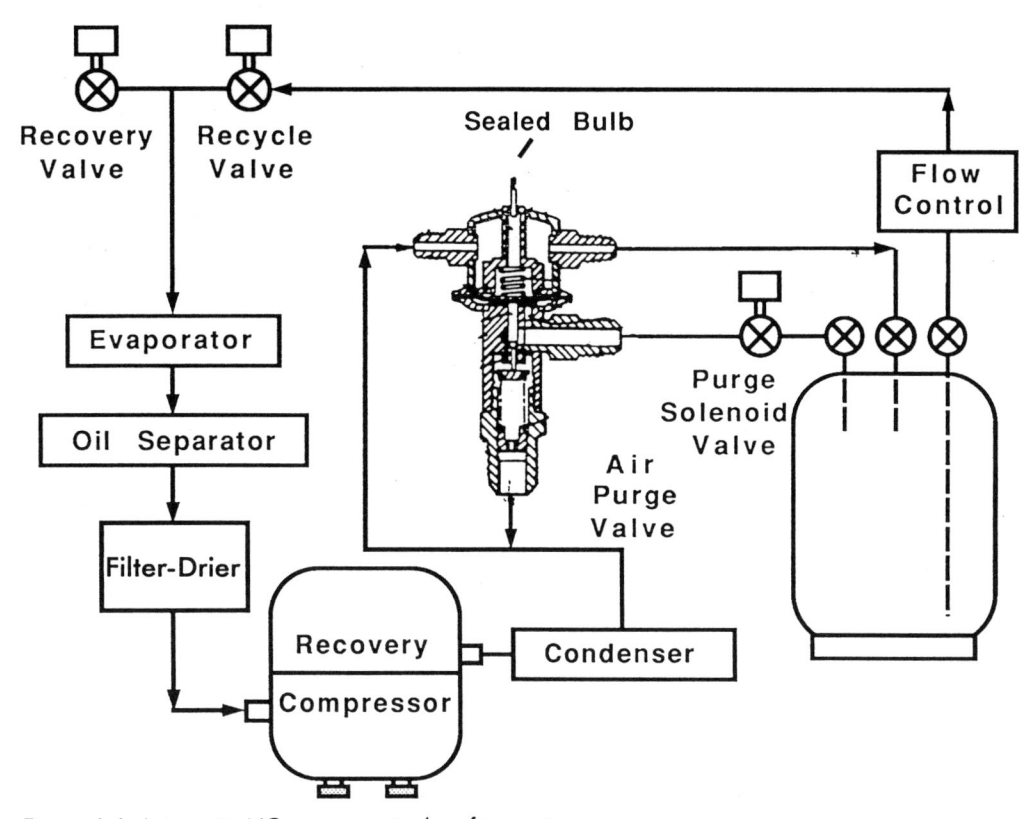

Figure 6-6. Automatic NC purge — single refrigerant

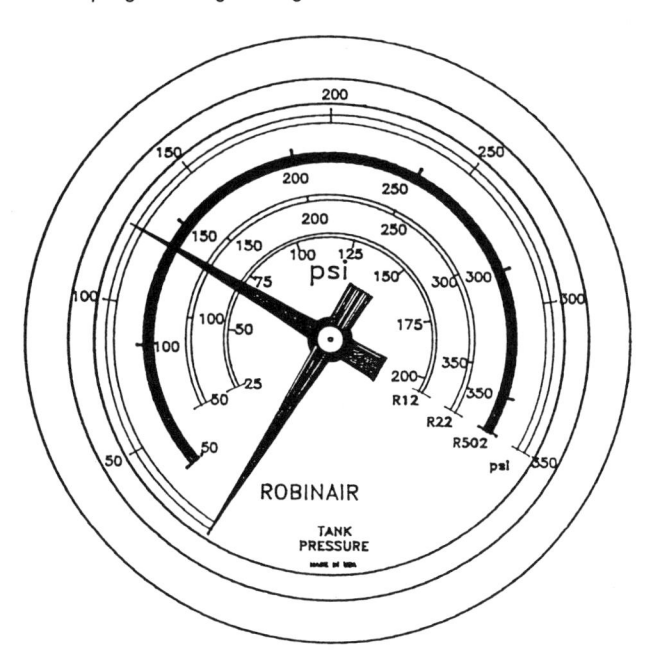

Figure 6-7. Dual-needle pressure gauge — multiple refrigerants

A differential temperature gauge[5] further improves NC measurement, Figure 6-8. One side of the gauge (gauge 1) connects to a sealed bulb containing R-12, while the other side connects to the top of the storage container. A second pressure gauge (gauge 2, not shown) calibrated to read in temperature, also connects to the sealed bulb.

The gauge 1 dial indicates an apparent saturation temperature for each refrigerant. The operator reads the actual temperature on gauge 2, then purges until the apparent temperature on gauge 1 for the particular refrigerant matches the temperature for the particular refrigerant.

This method has several advantages over the dual-needle pressure gauge. Gauge 1 has a smaller error due to both forces acting on a single diaphragm. At a given temperature, the difference between the tank pressure (R-22) compared to the reference (R-12) is smaller and less temperature dependent than either the R-22 or R-12 individual curves, Figure 6-9.

Different refrigerants fall in different angular sectors of the gauge 1 dial face, so it is usually possible to tell at a glance if the refrigerant is a different type than expected.

## Automatic Purge for Multiple Refrigerants
Partial pressure and oxygen sensor methods can be used.

## Electronic Partial-Pressure Purge
To find the partial pressure of air, the saturation pressure of the refrigerant is subtracted from the total pressure in the air-purge chamber.

The refrigerant saturation pressure may be directly measured in a sealed bulb at the same temperature, or it can be calculated from a temperature measurement. Using electronics, such calculations can easily be made for a large number of refrigerants.

A simple schematic[5,6] using a temperature probe and a pressure transducer both placed near the storage container is shown in Figure 6-10. The microprocessor calculates saturation pressure from the temperature and compares it to the measured pressure. If the partial pressure of the NCs exceeds the allowable level for that refrigerant, the NCs are purged.

Alternatively, two pressure transducers can be used. One measures pressure in a sealed bulb that may contain R-12, while the other measures pressure in the refrigerant line near the storage container.

Air and Non-Condensables

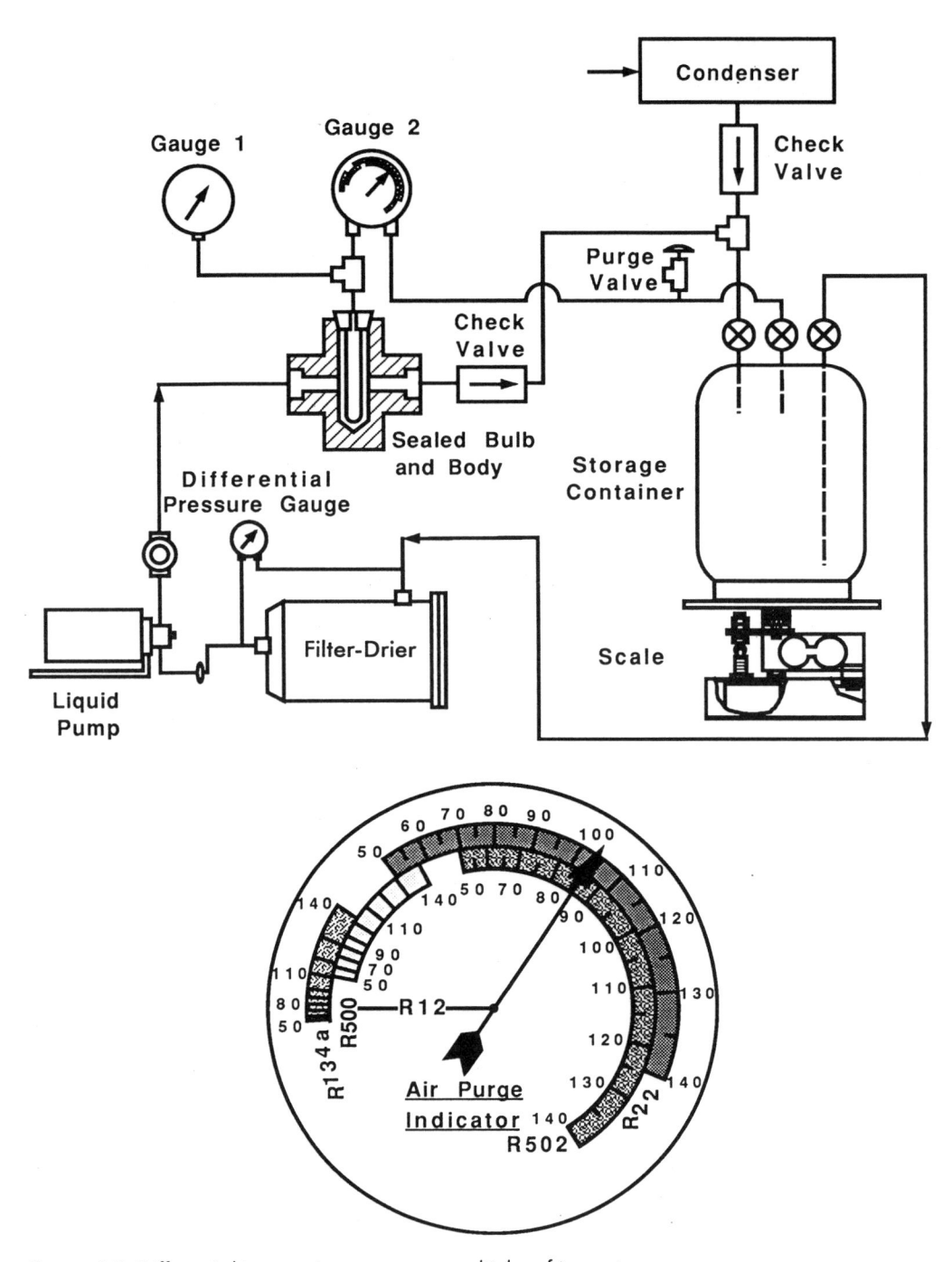

*Figure 6-8. Differential temperature gauge — multiple refrigerants*

81

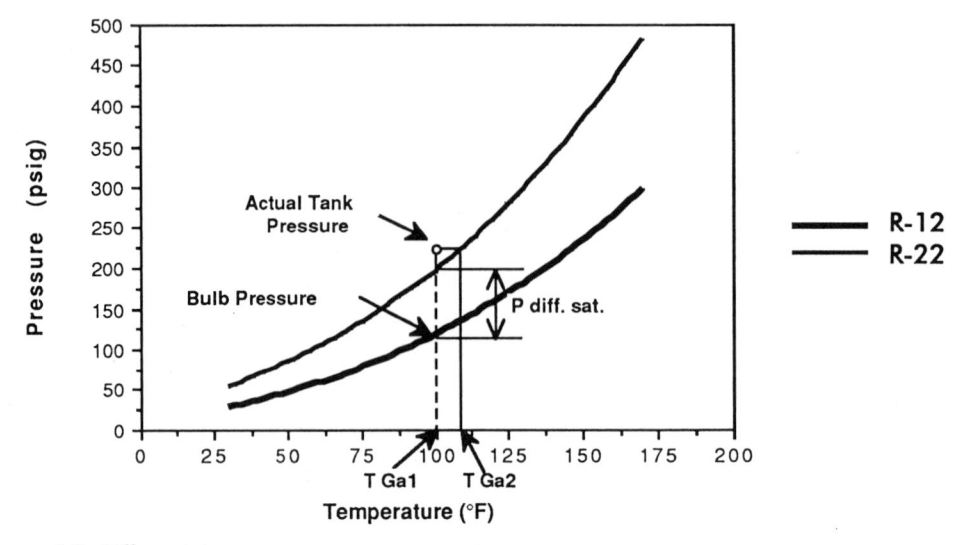

*Figure 6-9. Differential temperature gauge operation*

## Measuring NCs in a Storage Container

When checking a separate storage container for NCs, one simple method[7] is to use a thermometer and pressure gauge, Figure 6-11. The tank must be at room temperature, which may take up to 12 hours. A good time to check a tank is first thing in the morning.

Shaking the tank to reach temperature equilibrium may cause NCs to dissolve in the liquid. Using a saturation pressure-temperature chart, Figure 6-12, the need for NC purging can be determined. While it is not precise, this method may be used to detect and purge excessive NCs.

# Purging NCs When Recycling

NCs are most commonly trapped in the storage container. As shown in Figure 6-13, the refrigerant and NCs are pumped into the top of the container, where liquid refrigerant settles to the bottom and the NCs remain on top. A check valve prevents migration of NCs back into the recovery unit. Circulating the refrigerant from and to the tank further assists in separating dissolved NCs from the liquid refrigerant.

A single-pass air-separation system[8], Figure 6-14, uses a separate purge chamber located between the condenser and the storage container. A liquid trap keeps air from migrating out of the chamber and into the container. A solenoid and a float maintain the liquid level, so a relatively high NC concentration can be maintained in the purge chamber. Purging can be performed less frequently and more efficiently compared to storage container purging.

Air and Non-Condensables

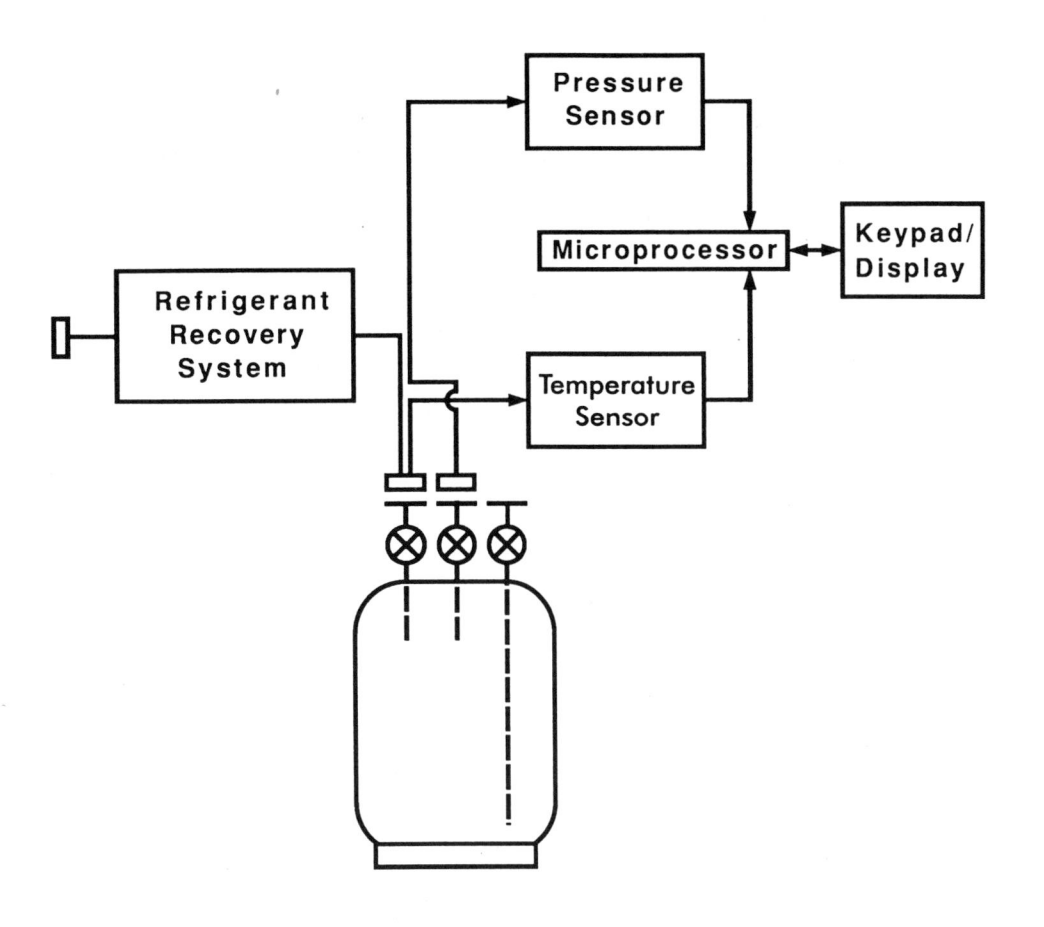

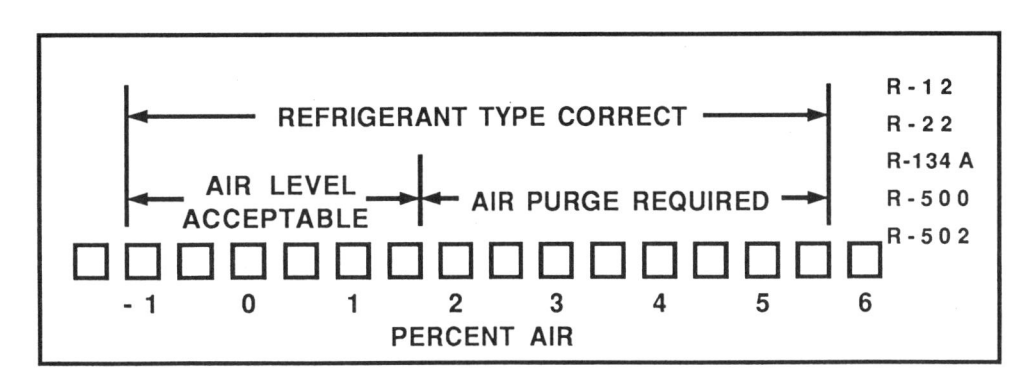

*Figure 6-10. Electronic NC purge — multiple refrigerants*

83

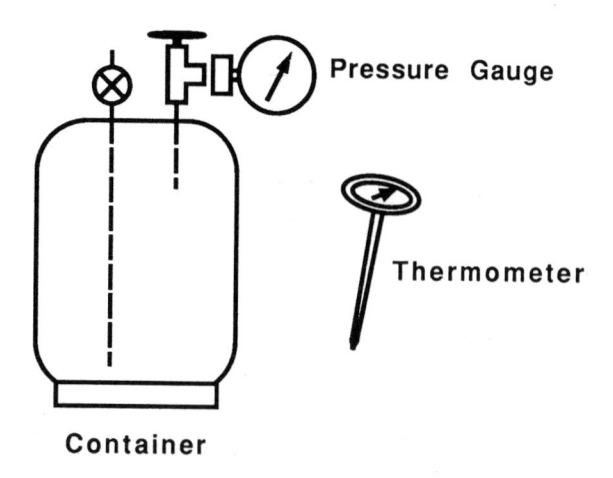

*Figure 6-11. Simple NC checking procedure*

**TEMPERATURE PRESSURE CHART** — Sporlan

Vacuum-Inches of Mercury — Italic Figures
Pressure-Pounds Per Square Inch Gage — Bold Figures

| TEMPERATURE °F. | 22-V | 502-R | 12-F | 134a-J | 717-A | TEMPERATURE °F. | 22-V | 502-R | 12-F | 134a-J | 717-A | TEMPERATURE °F. | 22-V | 502-R | 12-F | 134a-J | 717-A |
|---|---|---|---|---|---|---|---|---|---|---|---|---|---|---|---|---|---|
| −60 | 11.9 | 7.2 | 19.0 | 21.6 | 18.7 | 12 | 34.8 | 43.2 | 15.9 | 13.2 | 25.5 | 42 | 71.5 | 83.8 | 38.9 | 37.0 | 61.4 |
| −55 | 9.2 | 3.9 | 17.3 | 20.2 | 16.7 | 13 | 35.8 | 44.3 | 16.5 | 13.8 | 26.4 | 43 | 73.0 | 85.4 | 39.8 | 38.0 | 62.9 |
| −50 | 6.1 | 0.2 | 15.4 | 18.6 | 14.4 | 14 | 36.8 | 45.4 | 17.1 | 14.4 | 27.4 | 44 | 74.5 | 87.0 | 40.8 | 39.0 | 64.5 |
| −45 | 2.7 | 1.9 | 13.3 | 16.7 | 11.8 | 15 | 37.8 | 46.5 | 17.7 | 15.1 | 28.3 | 45 | 76.1 | 88.7 | 41.7 | 40.0 | 66.1 |
| −40 | 0.6 | 4.1 | 11.0 | 14.7 | 8.8 | 16 | 38.8 | 47.7 | 18.4 | 15.7 | 29.3 | 46 | 77.6 | 90.4 | 42.7 | 41.1 | 67.6 |
| −35 | 2.6 | 6.5 | 8.4 | 12.3 | 5.5 | 17 | 39.9 | 48.9 | 19.0 | 16.4 | 30.3 | 47 | 79.2 | 92.1 | 43.7 | 42.2 | 69.3 |
| −30 | 4.9 | 9.2 | 5.5 | 9.7 | 1.7 | 18 | 40.9 | 50.0 | 19.7 | 17.1 | 31.3 | 48 | 80.8 | 93.9 | 44.7 | 43.2 | 70.9 |
| −25 | 7.5 | 12.1 | 2.3 | 6.8 | 1.2 | 19 | 42.0 | 51.2 | 20.4 | 17.7 | 32.4 | 49 | 82.4 | 95.6 | 45.7 | 44.3 | 72.6 |
| −20 | 10.2 | 15.3 | 0.6 | 3.6 | 3.5 | 20 | 43.1 | 52.5 | 21.1 | 18.4 | 33.4 | 50 | 84.1 | 97.4 | 46.7 | 45.4 | 74.3 |
| −18 | 11.4 | 16.7 | 1.3 | 2.2 | 4.5 | 21 | 44.2 | 53.7 | 21.8 | 19.2 | 34.5 | 55 | 92.6 | 106.6 | 52.1 | 51.2 | 83.2 |
| −16 | 12.6 | 18.1 | 2.1 | 0.7 | 5.6 | 22 | 45.3 | 54.9 | 22.5 | 19.9 | 35.5 | 60 | 101.6 | 116.4 | 57.8 | 57.4 | 92.6 |
| −14 | 13.9 | 19.5 | 2.8 | 0.4 | 6.7 | 23 | 46.5 | 56.2 | 23.2 | 20.6 | 36.6 | 65 | 111.3 | 126.7 | 63.8 | 64.0 | 102.8 |
| −12 | 15.2 | 21.0 | 3.7 | 1.2 | 7.8 | 24 | 47.6 | 57.5 | 23.9 | 21.4 | 37.7 | 70 | 121.5 | 137.6 | 70.2 | 71.1 | 113.8 |
| −10 | 16.5 | 22.6 | 4.5 | 2.0 | 9.0 | 25 | 48.8 | 58.8 | 24.6 | 22.1 | 38.8 | 75 | 132.2 | 149.1 | 77.0 | 78.7 | 125.5 |
| −8 | 17.9 | 24.2 | 5.4 | 2.8 | 10.2 | 26 | 50.0 | 60.1 | 25.4 | 22.9 | 40.0 | 80 | 143.7 | 161.2 | 84.2 | 86.7 | 138.0 |
| −6 | 19.4 | 25.8 | 6.3 | 3.7 | 11.5 | 27 | 51.2 | 61.5 | 26.2 | 23.7 | 41.2 | 85 | 155.7 | 174.0 | 91.7 | 95.2 | 151.4 |
| −4 | 20.9 | 27.5 | 7.2 | 4.6 | 12.8 | 28 | 52.4 | 62.8 | 26.9 | 24.5 | 42.4 | 90 | 168.4 | 187.4 | 99.7 | 104.3 | 165.5 |
| −2 | 22.4 | 29.3 | 8.2 | 5.5 | 14.2 | 29 | 53.7 | 64.2 | 27.7 | 25.3 | 43.7 | 95 | 181.9 | 201.4 | 108.1 | 113.9 | 180.6 |
| 0 | 24.0 | 31.1 | 9.2 | 6.5 | 15.6 | 30 | 54.9 | 65.6 | 28.5 | 26.1 | 44.9 | 100 | 196.0 | 216.2 | 117.0 | 124.1 | 196.7 |
| 1 | 24.8 | 32.0 | 9.7 | 7.0 | 16.4 | 31 | 56.2 | 67.0 | 29.3 | 26.9 | 46.1 | 105 | 210.8 | 231.7 | 126.4 | 134.9 | 213.9 |
| 2 | 25.7 | 32.9 | 10.2 | 7.5 | 17.1 | 32 | 57.5 | 68.4 | 30.1 | 27.8 | 47.4 | 110 | 226.4 | 247.9 | 136.2 | 146.4 | 231.8 |
| 3 | 26.5 | 33.9 | 10.7 | 8.0 | 17.9 | 33 | 58.8 | 69.9 | 30.9 | 28.6 | 48.7 | 115 | 242.8 | 264.9 | 146.5 | 158.4 | 251.0 |
| 4 | 27.4 | 34.9 | 11.3 | 8.6 | 18.7 | 34 | 60.2 | 71.3 | 31.8 | 29.5 | 50.0 | 120 | 260.0 | 282.7 | 157.3 | 171.1 | 271.1 |
| 5 | 28.3 | 35.9 | 11.8 | 9.1 | 19.5 | 35 | 61.5 | 72.8 | 32.6 | 30.4 | 51.4 | 125 | 278.1 | 301.4 | 168.6 | 184.5 | 292.5 |
| 6 | 29.1 | 36.9 | 12.4 | 9.7 | 20.3 | 36 | 62.9 | 74.3 | 33.5 | 31.3 | 52.7 | 130 | 297.0 | 320.8 | 180.5 | 198.7 | 314.9 |
| 7 | 30.0 | 37.9 | 12.9 | 10.2 | 21.1 | 37 | 64.3 | 75.9 | 34.3 | 32.2 | 54.1 | 135 | 316.8 | 341.2 | 192.9 | 213.5 | 338.8 |
| 8 | 31.0 | 38.9 | 13.5 | 10.8 | 22.0 | 38 | 65.7 | 77.4 | 35.2 | 33.1 | 55.5 | 140 | 337.5 | 362.6 | 205.9 | 229.2 | 363.5 |
| 9 | 31.9 | 39.9 | 14.1 | 11.4 | 22.8 | 39 | 67.1 | 79.0 | 36.1 | 34.1 | 57.0 | 145 | 359.1 | 385.0 | 219.5 | 245.6 | 390.2 |
| 10 | 32.8 | 41.0 | 14.7 | 12.0 | 23.7 | 40 | 68.6 | 80.5 | 37.0 | 35.0 | 58.4 | 150 | 381.7 | 408.4 | 233.7 | 262.8 | 417.4 |
| 11 | 33.8 | 42.1 | 15.3 | 12.6 | 24.6 | 41 | 70.0 | 82.1 | 37.9 | 36.0 | 59.2 | 155 | 405.4 | 432.9 | 248.6 | 281.0 | 447.0 |

*Figure 6-12. Saturation pressure-temperature chart (Courtesy, Sporlan Valve Company)*

## Purging Efficiency

Refrigerant is also lost if NCs are purged by some current methods. For recycling equipment to meet SAE and ARI standards[9,10,11], current U.S. EPA regulations allow up to 5% of the refrigerant to be lost during air purge. ARI and ISO revised standards (ISO[12]), due to issue soon, are expected to allow only 3% refrigerant loss from NCs during purging, oil draining, and unit clearing — combined.

While very efficient purge units are used on low-pressure refrigeration (R-11, R-123) systems, these units are too expensive for recycling equipment. The following three improved methods are under development:

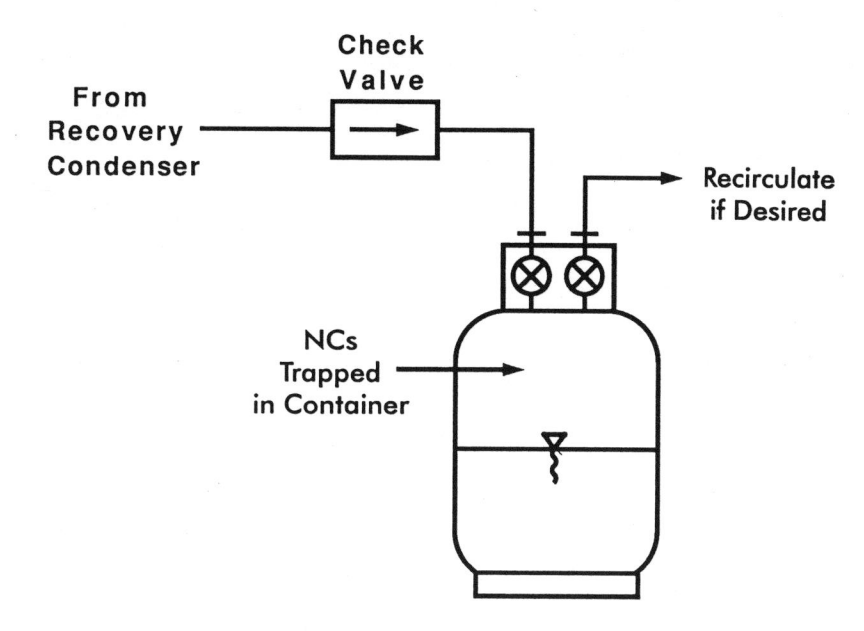

*Figure 6-13. Trapping NCs in storage container*

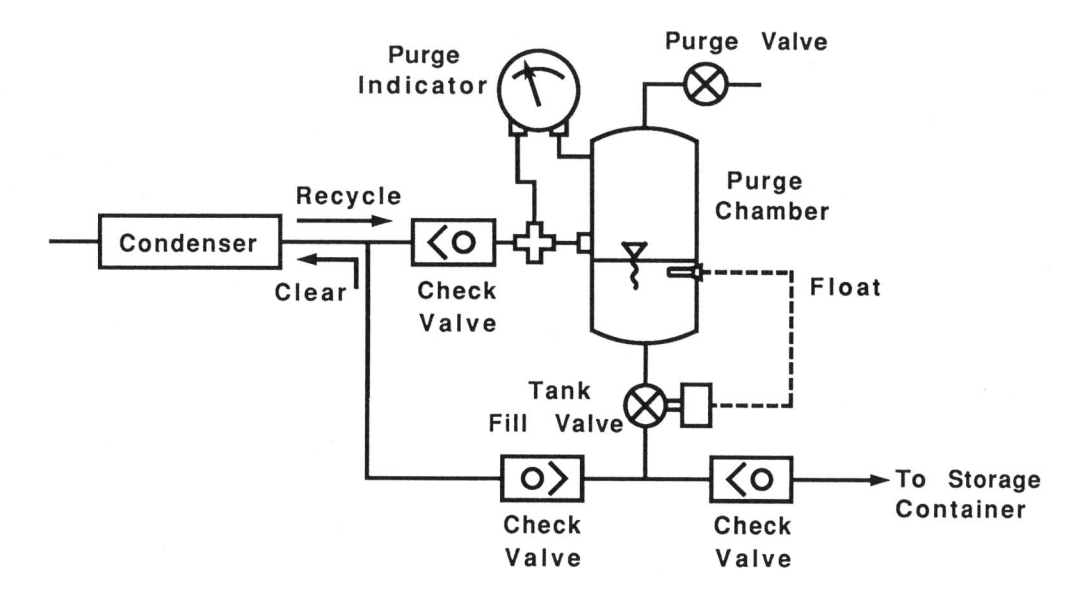

*Figure 6-14. Trapping NCs in purge chamber*

- A selective membrane allows either refrigerant or air to pass through and prevents or impedes the other.
- Refrigerant and NCs purged through a chemical material that adsorbs the refrigerant.
- Cooling condenses most of the refrigerant before purging.

## Desiccant Adsorption Method

An inexpensive but efficient air purge method[8] is shown in Figure 6-15. In this method, refrigerant-and-air vapor purges through a desiccant adsorption canister, where molecular sieve, activated charcoal, or other sorbent materials adsorb the refrigerant, and air passes through.

In a normal recovery-recycling sequence, refrigerant passes through the evaporator, compressor, refrigerant clearing valve port 2, condenser, check valve 1, air-purge chamber, tank-fill valve, and check valve 3 on the way to the storage container. When the purge indicator senses a need to purge, air purge valves 1 and 2 are opened for a period of time (determined by the purge indicator or some other means).

During this time, the heat exchanger tubes inside the desiccant adsorption canister are connected through refrigerant clearing valve ports 3 and 4 to the evaporator inlet, so the canister is relatively cool compared to the air-purge chamber. An orifice within or adjacent to air purge valve 1 controls the mass flow rate of the purge. (One method of determining when the desiccant adsorption canister is saturated is to monitor the time air-purge valve 1 is open; multiply by the NC-refrigerant purge flow rate to obtain pounds of refrigerant adsorbed.)

At the end of the recovery-recycling sequence, or when the adsorption canister is saturated, the recovery-recycling valve closes and the refrigerant clearing valve switches to a clearing sequence, whereby port 1 is connected to port 4 to pump hot refrigerant vapor through the heat exchanger tubes of the adsorption canister and check valve 4, to the storage container. Meanwhile, air-purge valve 2 and the tank-fill valve open. Ports 2 and 3 of the refrigerant clearing valve are connected so the compressor removes refrigerant from the condenser, purge chamber, and adsorption canister.

The combination of heating the canister and the vacuum provided by the compressor suction drives the adsorbed refrigerant out of the desiccant. The regenerated canister is then ready for the next purge operation.

## Sampling for NC Analysis

It is difficult to know where to draw a sample from a refrigeration system for NC analysis. In locations where air is trapped, high concentrations would be measured. In other locations, very low concentrations may be found. The NCs in a running system may be trapped at the top of the condenser or receiver, then when the system shuts down, the NCs migrate.

Samples for NC analysis are drawn from the vapor space. Use the sampling container and procedure recommended or furnished by a competent laboratory that is experienced in refrigerant analysis. The laboratory will perform NC analysis using gas chromatography.

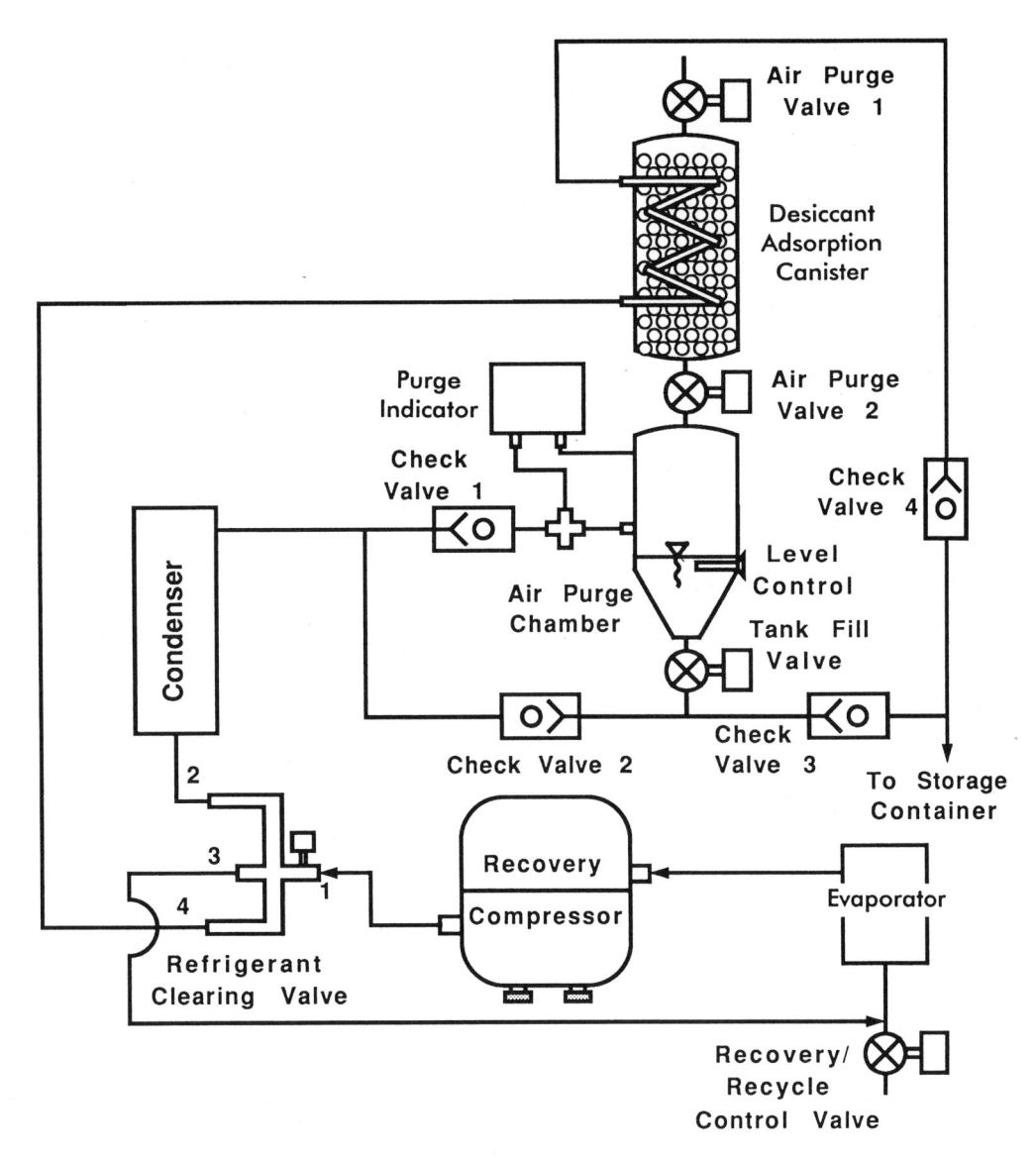

*Figure 6-15. Efficient NC purge using desiccant adsorption*

# Summary

Air is a serious contaminant in refrigeration systems. Proper evacuation is essential to remove air and moisture from systems. Low-permeation valved hoses reduce the amount of air introduced while making service connections.

Non-condensables (NCs) are measured using partial-pressure methods or oxygen sensors. In recycling units, manual and automatic methods can be used for both single and multiple refrigerants.

Air may be trapped in the storage container or in a separate chamber. Increased emphasis on recycling unit purging efficiency is expected; the desiccant adsorption method looks promising.

It is difficult to know where to draw a sample from a refrigeration system for NC analysis. A vapor sample from a storage container can be sent to a competent laboratory for analysis using gas chromatography.

# References

[1] ASHRAE, *1994 Handbook*, Chapter 6.

[2] U.S. Patent 5,005,369, Robinair Division, SPX Corporation, 1991.

[3] U.S. Patent 5,065,595, Sporlan Valve Company, 1991.

[4] U.S. Patent 5,063,749, Robinair Division, SPX Corporation, 1991.

[5] U.S. Patent 5,181,391, Robinair Division, SPX Corporation, 1993.

[6] U.S. Patent 5,285,647, Robinair Division, SPX Corporation, 1994.

[7] SAE J1989, 1989, *Recommended Service Procedure for the Containment of R-12*.

[8] U.S. Patent Pending, Robinair Division, SPX Corporation.

[9] SAE J1990, 1989, *Extraction and Recycle Equipment for Mobile Automotive Air-Conditioning Systems*.

[10] SAE J2210, 1991, *HFC-134a Recycling Equipment for Mobile Air-Conditioning Systems*.

[11] ARI Standard 740-93, 1993, *For Performance of Refrigerant Recovery and/or Recycling Equipment*.

[12] ISO/WI 1165OR, *Performance of Refrigerant Recovery and/or Recycling Equipment*, 1994.

# Chapter Seven

# Mixed Refrigerants

A mixed refrigerant is any refrigerant that still would not meet product specifications even if all moisture, acid, particulates, oil, and non-condensables were removed. For pure refrigerants or blends, one or more "other" refrigerants may be present. For blends, the proportion of the component refrigerants may have changed.

A blend is any refrigerant that is made up of two or more component refrigerants that does not act as a true mixture. Also referred to as zeotropes, blends have the following in common:

- They do not have the same composition in the liquid and vapor phase of a container.
- They do not have a single boiling point, but rather a boiling point range. This characteristic is often referred to as "glide."
- Their component refrigerants permeate through a hose or leak through a joint at different rates.

## Causes of Mixed Refrigerants

While mixed refrigerant can be the result of chemical reactions or selective leakage, faulty service practices and refrigerant handling procedures are normally responsible.

Refrigerants are remarkably stable compounds that last indefinitely under normal operating conditions. At extremely high temperatures, such as during a hermetic motor burnout, some of the refrigerant breaks down forming other refrigerants. Under high moisture conditions with air present, acids form that also break down some of the refrigerant. When chemical reaction occurs, it can often be traced to poor installation or service procedures.

Mixed refrigerants most often result from improper service procedures. Inadvertent mixing may be from failure to:

- dedicate and clearly mark containers for specific refrigerants;

- clear hoses or recovery equipment before switching to a different refrigerant;
- test suspect refrigerant before consolidating into large batches;
- use proper retrofit procedures.

Blends are subject to selective leakage of the component refrigerants. The refrigerant composition may be different at the beginning of recovery than at the end, depending on where the service ports are located.

Mixed refrigerants may affect the following system characteristics:

- Performance, operating characteristics, capacity, and efficiency of the system
- Material compatibility, lubrication, equipment life, and warranty costs
- Increased refrigerant emissions through air purge devices and leakage through hoses or joints
- Increased service and repair requirements and higher operating costs

For example, mixed refrigerant can cause severe damage to motor vehicle air conditioners. When R-134a and PAG vehicles contain more than 3% R-12, the PAG lubricant reacts with the R-12. While developing the SAE R-12 to R-134a retrofit procedure[1], sealed-tube tests were used to determine the PAG lubricant tolerance for R-12. Glass tubes containing steel coupons, copper coupons, PAG oil, as well as varying amounts of R-12 and R-134a, were prepared and then aged 14 days at 300°F (177°C). Tubes with more than 3% R-12 showed increasing reactivity with increasing R-12 concentration. In Figure 7-1, the tube on the right containing 100% R-12, turned black and produced floating solid chunks, while the tube on the left containing no R-12 showed no visible reaction.

## Protecting the Used Refrigerant Supply

The used refrigerant supply is the sum of refrigerant in existing systems and what has been collected for recycling, reclaiming, and storage. When new production ceases, this becomes the only source of that refrigerant for service. Under the Montreal Protocol, the used refrigerant supply will be the only source of CFC refrigerants after 1995. Given this situation, major concerns are:

- inability or high cost of separating refrigerants;
- high cost of disposal and loss of refrigerant for future service.

If one container of mixed refrigerant is consolidated with pure refrigerant, the whole batch becomes contaminated. Chemicals added to a single system eventually enter the used refrigerant supply. While it may take some time and many occurrences to taint the very large used refrigerant supply, the economic impact if it becomes contaminated will be staggering.

## Mixed Refrigerants

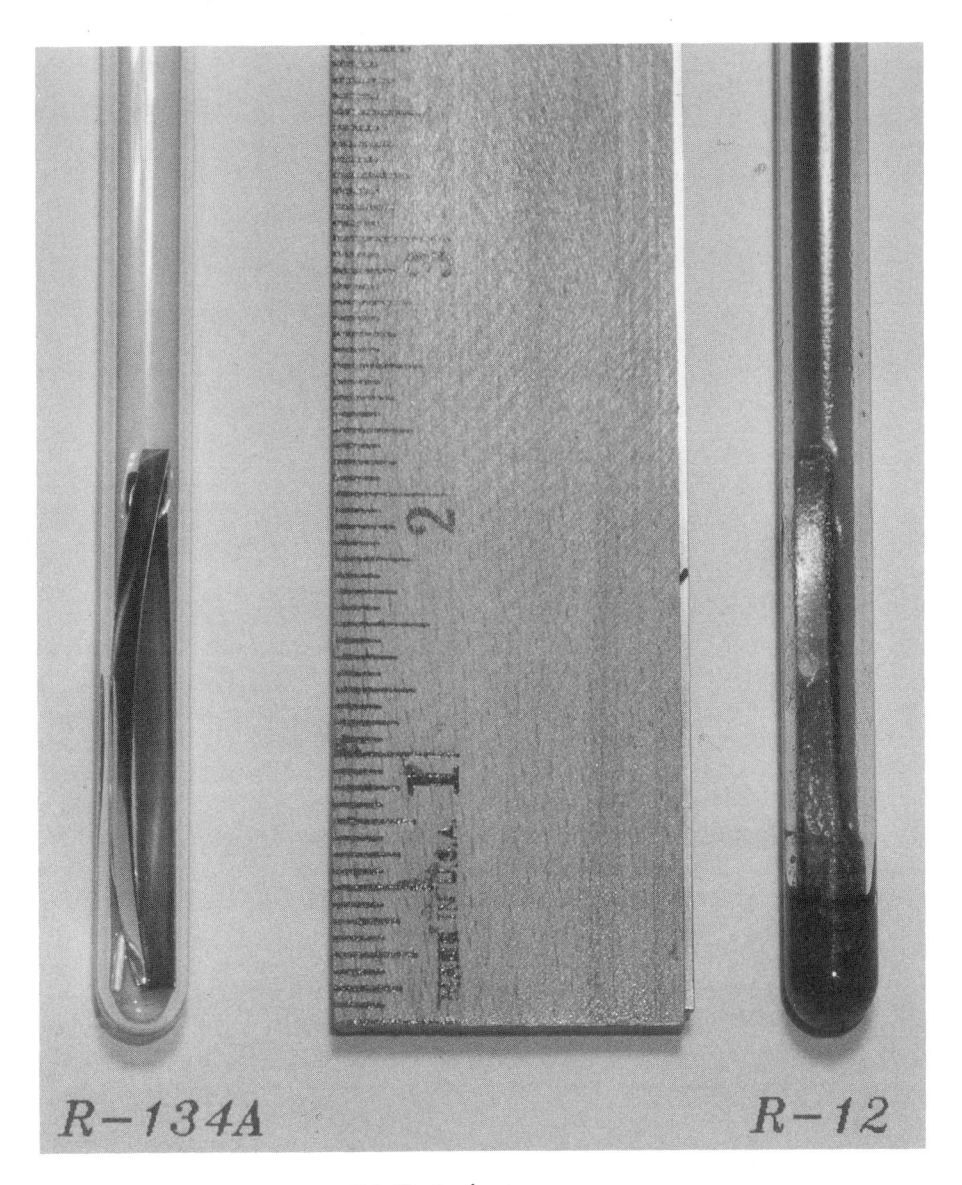

### PAG Lubricant
### Copper, Aluminum, Steel

*Figure 7-1. Sealed tubes with R-12 and R-134a/PAG (Aged 14 days at 300°F) (Courtesy, Robinair Division, SPX Corporation)*

Individual actions are required by all to prevent such an occurrence. There are two opportunities to protect the used refrigerant supply. The first is to avoid service procedures that create mixed refrigerant. The second is to identify suspected mixed refrigerant and isolate it from the used refrigerant supply until it can be tested.

## Selling Used Refrigerant

Mixed refrigerant is a big issue when selling used refrigerants, especially when using "arm's length" transactions such as between contractors. If a contractor knows where the refrigerant came from, how it was processed, and which system it is to be charged to, the risk is lessened while the responsibility is clear.

While some draw the line at a change of ownership, the owner often relies on the contractor to make such decisions. It is good practice to inform the new owner when charging with recycled refrigerant. When the same-contractor chain of responsibility is broken, refrigerant should be analyzed and verified to meet new product specifications before being sold. (For more information, obtain the pamphlet *Handling and Reuse of Refrigerants in the United States* from the Air-Conditioning and Refrigeration Institute[2].)

## Preventing Mixed Refrigerant

One of the best methods of preventing mixed refrigerant is to clear recycling equipment (see Chapter 3). Clearing the recycling equipment involves transferring as much refrigerant as possible into the storage container, draining oil from the suction oil separator, then pulling a deep vacuum on the equipment. The advantages of an automatic clearing sequence, Figure 7-2, were also discussed in Chapter 3.

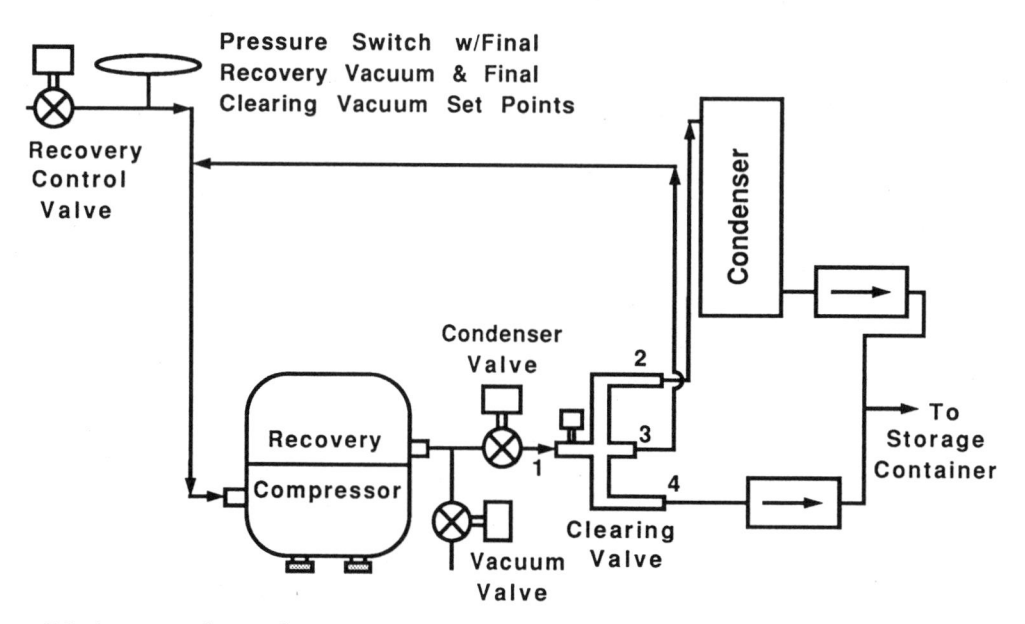

*Figure 7-2. Automatic clearing for recovery-recycling equipment*

# Mixed Refrigerants

In a clearing mode, when ports 1 and 4 and ports 2 and 3 are connected, the compressor pumps through the condenser valve until the pressure reaches the final recovery vacuum (15 in. Hg, or 371 mm), then pumps through the vacuum valve until the pressure reaches the final clearing vacuum (29 in. Hg, or 737 mm). Two automatic methods that minimize operator error are:

- Clearing after each recovery operation — After each recovery cycle, the recovery control valve automatically closes and the reversing valve switches to the clearing position while the compressor continues to run. When the low-pressure switch senses clearing is complete, the compressor shuts down and the clearing valve reverts to the recovery position. Finally, the oil separator is drained and a separate vacuum pump completes refrigerant removal.
- Clearing only when switching refrigerants — After each recovery cycle, drain the oil with the compressor off. When changing refrigerants, an electrical clear switch is activated. The compressor and clearing valve operate as previously described. Electronic controls automatically switch the reversing valve to the recovery mode in preparation for the next recovery.

The efficiency of clearing procedures has been determined[3]. In one experiment, 10 lb (4.5 kg) of refrigerant were recovered 20 times for each of R-12, R-22, and R-502, switching back and forth at random. After using the procedure described in Chapter 3, a five minute vacuum was pulled using a 1.2 cfm (0.57 l/sec) rotary vane vacuum pump on the recovery unit. At the end of the experiment, the containers were analyzed by gas chromatography, producing the results in Table 7-1.

The three tanks (R-12-1, R-12-2, and R-12-3) represent the main recovery tank and the first and second evacuated tanks for the same batch. The R-12 average purity was 95.13%, with each of the 20 runs contributing 0.22%. The R-22 average purity was 93.7% with each run contributing 0.31%. Most of the other refrigerant was R-502, which contains R-22. The R-502 average purity was 98.78% with each run contributing 0.07%.

While not practical for all refrigeration industry sectors, dedicating service equipment to a single refrigerant effectively prevents mixed refrigerant.

## Motor Vehicle Air Conditioners

Vehicle manufacturers have protected the used refrigerant supply from mixing with other refrigerants by limiting the number of approved refrigerants to just R-12 and R-134a for service, retrofit, and new vehicles. In addition, unique vehicle service ports are provided for R-134a vehicles.

Within the framework of the SAE Defrost and Interior Climate Committee, vehicle manufacturers, chemical producers, component suppliers, recycling equipment manufacturers, and aftermarket service associations wrote SAE standards further guarding against mixing by requiring separate service fittings[4] for R-134a hoses,

|  | Wt %<br>R-502 | Wt %<br>R-22 | Wt %<br>R-12 | Wt %<br>Total | Change in Wt %<br>Per Run * |
|---|---|---|---|---|---|
| New R-12 | 0.44 | 0.08 | 99.48 | 100.00 | .00 |
| R-12-1 | 1.52 | 2.43 | 96.05 | 100.00 | .17 |
| R-12-2 | 1.76 | 3.15 | 95.09 | 100.00 | .22 |
| R-12-3 | 1.27 | 4.50 | 94.24 | 100.00 | .26 |
| New R-22 | 0.0 | 100.00 | 0.0 | 100.00 | .00 |
| R-22-1 | 4.06 | 93.35 | 2.60 | 100.00 | .33 |
| R-22-2 | 4.44 | 94.06 | 1.50 | 100.00 | .30 |
| New R-502 | 99.95 | 0.05 | 0.0 | 100.00 | .00 |
| R-502-1 | 98.84 | 0.0 | 1.16 | 100.00 | .06 |
| R-502-2 | 98.66 | 0.0 | 1.34 | 100.00 | .07 |
| R-502-3 | 98.83 | 0.01 | 1.16 | 100.00 | .07 |

\* Total change in weight percent divided by 20 runs.

*Table 7-1. Mixed refrigerant condition after 20 clearing operations*

storage containers, manifolds, vacuum pumps, charging equipment, and recycling equipment. Separate recycling equipment for R-12 and R-134a with appropriate hoses furnished is also required.

## Other Sectors

Household appliances and residential air conditioning systems may also benefit from dedicating recovery-recycling equipment to a single refrigerant. Nearly all residential air conditioners and heat pumps operating today use R-22. Contractors, having multiple service vehicles and mainly working on residential and light commercial air conditioners, may wish to dedicate several recovery-recycling units and storage containers for R-22 and to designate a few units for multiple refrigerants.

Almost all household refrigerators and freezers operating today use R-12. Many appliance manufacturers have selected R-134a for new refrigerators and recommend separate service equipment for each.

## Analyzing for Mixed Refrigerant

Refrigerant may be analyzed when both liquid and vapor phases are present in the sample container, such as in a storage tank. This is called a two-phase sample. Saturation vapor pressure and boiling point may be measured as described in the following paragraphs.

### Vapor Pressure

For any refrigerant, the vapor pressure in a two-phase container is a unique function of temperature. In Figure 7-3, the curves for R-12, R-22, and R-502 are unique and distinguishable throughout the expected temperature range. By contrast, R-12 and R-134a are not unique from each other from 40° to 90°F (4° to 32°C).

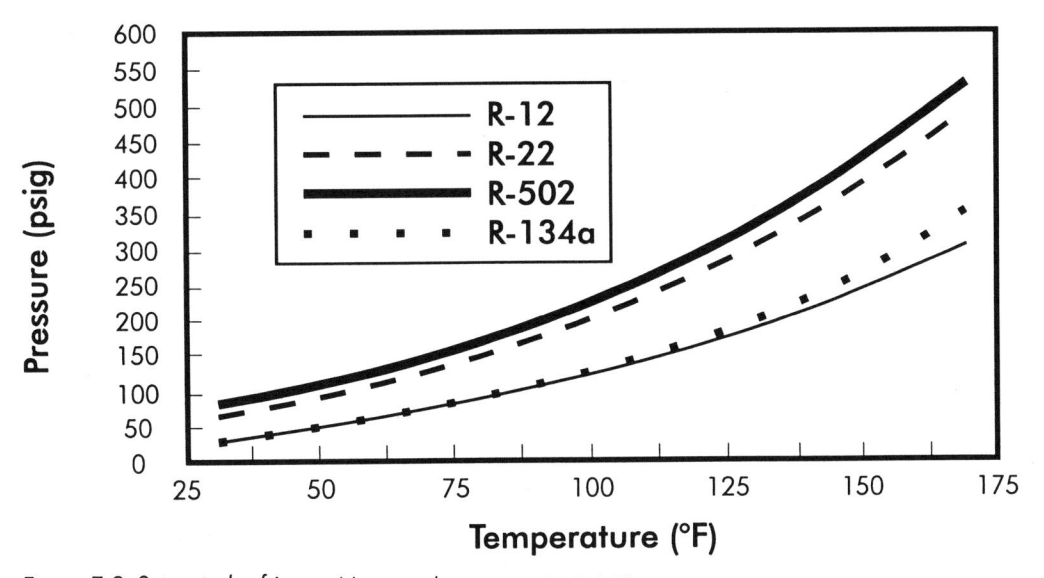

*Figure 7-3. Saturated refrigerant temperature-pressure curves*

When a refrigerant contains air, the pressure is higher than for the pure refrigerant. In previous chapters, the primary focus was on the partial pressure due to air; for this purpose, the primary focus is on the saturation pressure to determine the refrigerant type. The effect of air can be eliminated by purging air from the sample, calculating an expected range of air pressure, or directly measuring oxygen.

Vapor pressure measurements usually can identify between pure refrigerants but cannot differentiate mixed refrigerants with more than 90% purity from pure refrigerants. (Mixed refrigerants fall outside the expected pure refrigerant range.) The measurements required are as follows:

1. Pressure gauge-thermometer — read the pressure from the gauge and the temperature from the thermometer; compare these values to a chart like the

one in Figure 6-12. The pressure should be equal to or slightly higher (if air is present) than the chart value for the refrigerant.
2. Dual-needle pressure gauge[5] — read the tank pressure on the long needle on the outer scale. This pressure should equal or slightly exceed the pressure on one of the inner scales, thus identifying the correct refrigerant type (see Figure 6-7).
3. Differential temperature gauge[6] — the temperature on gauge 2 should be equal to or slightly less than the temperature on the appropriate refrigerant scale of gauge 1 to correctly identify the refrigerant (see Figure 6-8).
4. Electronic pressure-temperature indicator[7] — the values of pressure and temperature from the sensors are fed into a microprocessor that holds information on the refrigerant saturation curves shown in Figure 7-3. Beginning at the lowest curve (R-12), the microprocessor successively compares a calculated saturation pressure at the temperature for various refrigerants until it finds one that is equal to or slightly less than the measured pressure.

## Boiling Point

The boiling point is the temperature within a closed vessel containing two-phase refrigerant that will produce a pressure of one atmosphere (14.7 psia, or 101.3 kPa), Table 7-2.

| R-113  | 117.6°F | R-22   | - 41.4  |
|--------|---------|--------|---------|
| R-141b | 90.0    | R-502  | - 49.8  |
| R-123  | 82.6    | R-507  | - 51.7  |
| R-11   | 74.9    | R-143a | - 52.6  |
| R-114  | 38.8    | R-402a | - 53.3  |
| R-124  | 12.2    | R-125  | - 55.3  |
| R-134a | - 15.1  | R-402b | - 56.5  |
| R-12   | - 21.6  | AZ20   | - 62.5  |
| R-401a | - 27.6  | --     | --      |
| R-500  | - 28.3  | R-13   | -114.6  |
| R-401b | - 30.4  | R-23   | -115.7  |
|        |         | R-503  | -127.6  |

Table 7-2. Refrigerant boiling points

One method of determining the boiling point[8] is shown in Figure 7-4. Refrigerant fills the sample container to the liquid float level, which may be accomplished by drawing liquid into a vacuum or by cooling the sample container. Vapor refrigerant purges off the top of the container until the pressure reaches one atmosphere, at which time the boiling point temperature is recorded. The purged refrigerant is preferably captured into a container rather than vented to the atmosphere.

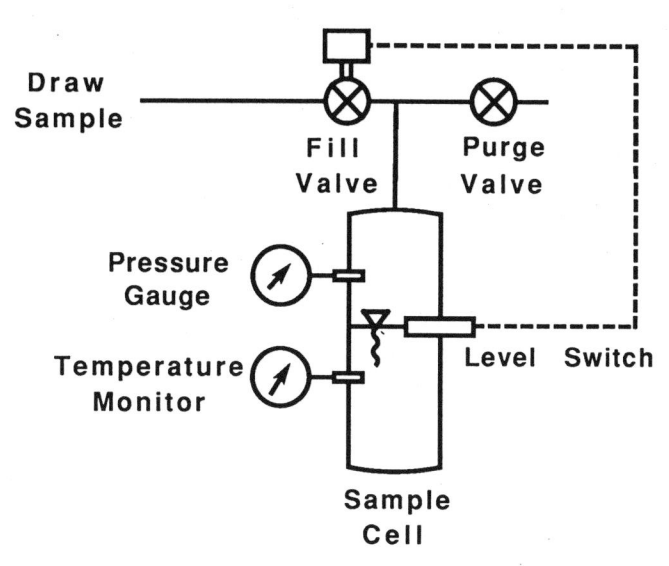

*Figure 7-4. Boiling point with internal probe*

A second method, shown in Figure 7-5, draws liquid refrigerant through a restricting orifice and directly measures the boiling point[8] as the refrigerant boils at the exit. When the refrigerant is captured in a previously evacuated container, it is necessary to measure the backpressure in the container to measure the boiling point temperature at the time when the pressure reaches 1 atmosphere.

Most refrigerants are distinguishable by their boiling point (see Table 7-2). While it may be helpful for an instrument to display the nearest pure refrigerant, merely displaying the boiling point provides useful information.

## Analyzing a Vapor Sample

Refrigerants can be distinguished by their gaseous properties. Apparatus[9] such as that shown in Figure 7-6 can be used to obtain a uniform vapor sample. The sample container is previously evacuated or purged to sweep out the previous sample and to lower the pressure. Liquid or vapor refrigerant, having a pressure higher than the final desired sample pressure, slowly feeds through the orifice or other restrictor. When the pressure reaches the final desired sample pressure, a valve is closed to stop filling and measure the desired property.

# The Challenge of Recycling Refrigerants

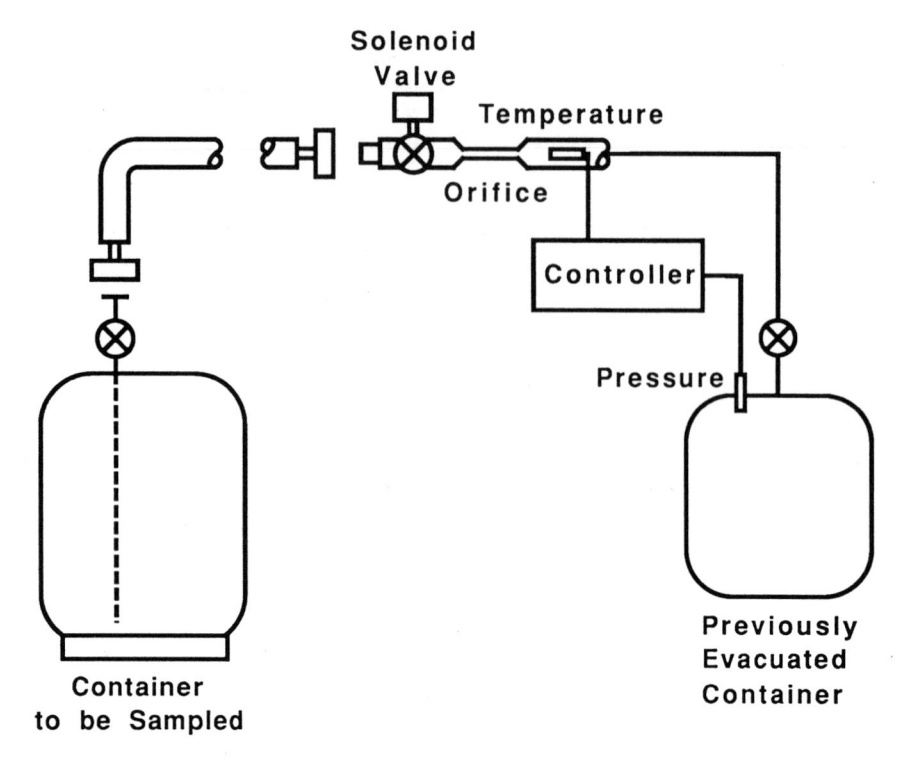

*Figure 7-5. Boiling point with spray probe*

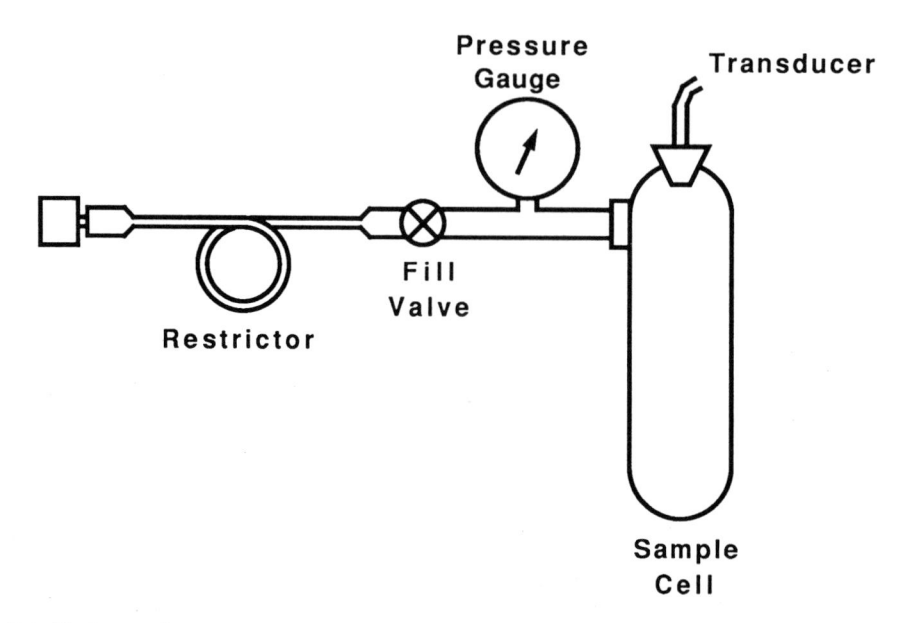

*Figure 7-6. Uniform refrigerant vapor-sampling apparatus*

If the measured gaseous property is temperature sensitive, it may be necessary to insulate, heat, or cool the sample chamber. Alternatively, the temperature can be measured and the device can compensate for the temperature-dependent portion of the measured property.

Some of the gaseous properties that may be used to distinguish refrigerants are as follows:

- Thermal conductivity as measured with a thermistor
- Dielectric constant, by measuring leakage current between parallel plates
- Molecular weight, by measuring wavelength of acoustic waves passing through the sample
- Light absorption, by measuring absorption of infrared light in various wavelength bands
- Gas chromatography

## Analyzing a Liquid Sample

Refrigerants can also be distinguished by their liquid properties. A sample container is cooled to recondense vapor, Figure 7-7, or filled by evacuation or purging to transfer liquid, Figure 7-8.

When recondensing vapor[7], an electric cooling strip lowers the sample cell temperature while the heat radiates to the atmosphere through the heat rejector. The refrigerant passes through a long passage in the sample cell to precool[8] before entering the central volume. The purge valve is used to fill the cell when non-condensables are present.

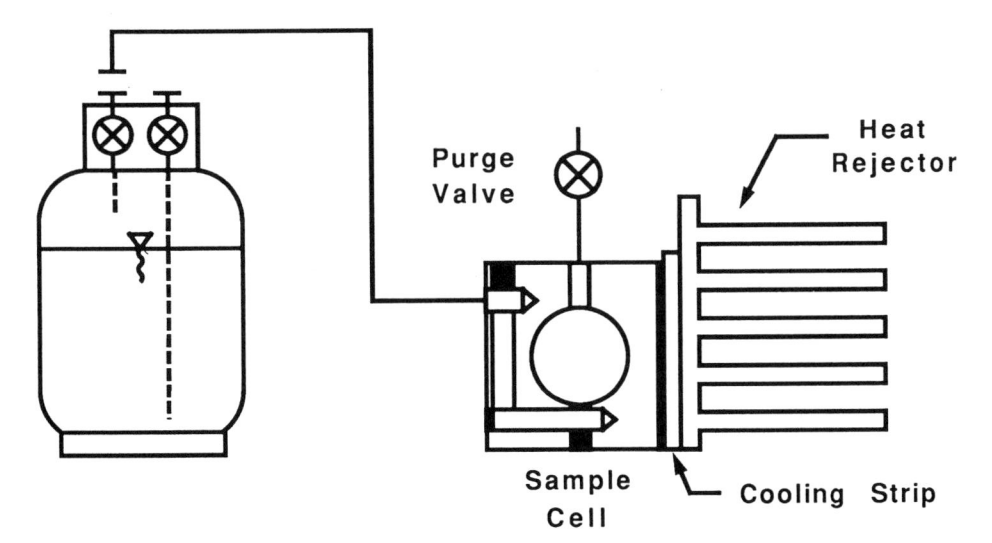

*Figure 7-7. Liquid sampling by recondensing vapor*

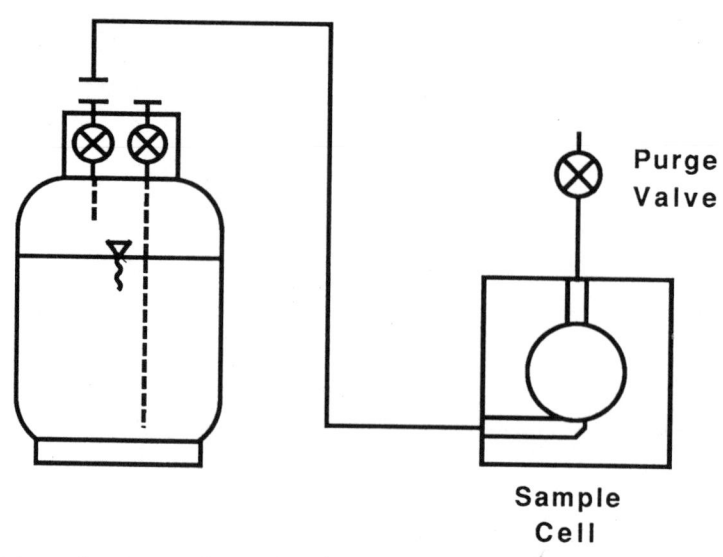

*Figure 7-8. Liquid sampling by transferring liquid*

Recondensing vapor keeps the sample cell clean of oil and particulates. However, when analyzing blends, the recondensed vapor sample is less consistent than a liquid sample transferred directly from the container. At additional expense, liquid refrigerant can be transferred into an accumulator and then reboiled and recondensed in a sample cell.

Transferring liquid refrigerant allows oil and particulates into the sample cell but provides a more representative sample for analyzing blends. Oils and particulates may affect liquid sample analysis, and the cell may require frequent cleaning.

Some of the liquid properties that may be used to distinguish refrigerants are as follows:

- Light absorption, by measuring absorption of infrared or near infrared light in various wavelength bands
- Electrical conductivity
- Thermal conductivity

## Analysis and Field Screening

Laboratory analysis most accurately determines whether refrigerant is pure or mixed. It is recommended to select ARI-certified reclaimers (or equivalent) because of their experience in refrigerant analysis.

Collect liquid refrigerant using sample containers and methods approved by the laboratory. The laboratory will analyze the refrigerant using gas chromatography, as called for in ARI Standard 740-93, and report the results.

The pamphlet *Handling and Reuse of Refrigerants in the United States*[2] sets a 98% refrigerant purity level. Refrigerant should contain less than 2% of "other refrigerants and volatile impurities" to be reused in a different system for the same owner.

Blends must also meet the chemical producer's tolerances for percentage of each component refrigerant. Blends are often similar to other blends or pure refrigerants. The 98% purity should be considered on an equivalent basis. For example, the mixture of 90% Du Pont "Suva" MP39 and 10% MP66 still meets MP39 specifications. Table 7-3 is a "98% equivalent" chart for several refrigerants. Field instruments capable of identifying "98% equivalent" pure refrigerants using infrared techniques will likely be introduced into automotive service in 1995.

Recycling equipment cannot unscramble mixed refrigerants. Refrigerant identifiers can screen refrigerant before recycling to avoid recycling costs when mixed refrigerant will be sent to a certified reclaimer.

While not capable of 98% equivalent determination, the following screening methods can identify extreme or likely mixed refrigerant conditions:

- Review equipment service history and reason for current service.
- Look for multiple refrigerants and storage containers on the same jobsite.
- Compare pressure and temperature measurements to expected values for the refrigerant.
- Measure other properties, such as boiling point for more precise screening.

## Summary

Mixed refrigerants not meeting new product specifications because of excessive "other refrigerants and other volatile impurities," contaminate the used refrigerant supply and may affect system performance. The condition is caused by faulty service procedures, system operation, or chemical reactions.

Opportunities to protect the used refrigerant supply include avoiding service procedures that cause mixed refrigerant and isolating suspected mixed refrigerant until it can be tested. Clearing recovery-recycling equipment, dedicating service equipment to a single refrigerant, using proper hoses, and labeling storage containers help prevent mixed refrigerants.

Vapor pressure, boiling point, thermal conductivity, and infrared absorption are refrigerant properties used to identify pure or mixed refrigerants. Field verification involves sampling for laboratory analysis (only method for reclaim quality) and screening methods for identifying suspected mixed refrigerant.

# The Challenge of Recycling Refrigerants

Base Refrigerant

| Contaminant | R-11 | R-12 | R-22 | R-114 | R-123 | R-125 | R-134a | MP39 R-401A | MP66 | HP80 R-402A R-402B | HP81 R-402B | HP62 FX70 R-404 | AZ50 R-507 | AC9000 | AZ20 | R-500 | R-502 |
|---|---|---|---|---|---|---|---|---|---|---|---|---|---|---|---|---|---|
| R-11 | X | 98 | 98 | 98 | 98 | 97 | 98 | 97 | 97 | 97 | 97 | 98 | 98 | 97 | 97 | 98 | 97 |
| R-12 | 98 | X | 98 | 98 | 98 | 98 | 98 | 97 | 97 | 97 | 97 | 98 | 98 | 97 | 97 | 93 | 98 |
| R-22 | 98 | 98 | X | 98 | 98 | 98 | 98 | 95 | 94 | 96 | 95 | 98 | 98 | 98 | 98 | 98 | 96 |
| R-114 | 98 | 98 | 98 | X | 98 | 97 | 98 | 97 | 97 | 97 | 97 | 98 | 97 | 97 | 97 | 98 | 97 |
| R-123 | 98 | 98 | 98 | 98 | X | 97 | 98 | 97 | 97 | 97 | 97 | 97 | 97 | 97 | 97 | 98 | 97 |
| R-125 | 97 | 97 | 98 | 97 | 97 | X | 98 | 97 | 97 | 94 | 96 | 96 | 96 | 95 | 95 | 97 | 98 |
| R-134a | 98 | 98 | 98 | 98 | 98 | 98 | X | 97 | 97 | 97 | 97 | 98 | 98 | 97 | 97 | 98 | 98 |
| R-401A | 98 | 98 | 98 | 98 | 98 | 98 | 98 | X | 80 | 96 | 95 | 97 | 97 | 97 | 97 | 97 | 95 |
| MP66 | 98 | 98 | 96 | 98 | 98 | 98 | 98 | 80 | X | 95 | 94 | 96 | 95 | 97 | 96 | 97 | 95 |
| R-402A | 97 | 98 | 95 | 97 | 97 | 95 | 98 | 96 | 96 | X | 92 | 96 | 96 | 96 | 96 | 97 | 94 |
| R-402B | 97 | 98 | 97 | 97 | 97 | 97 | 98 | 95 | 94 | 92 | X | X | 80 | 95 | 95 | 97 | 97 |
| R-404 | 97 | 98 | 98 | 97 | 97 | 96 | 98 | 97 | 97 | 96 | 96 | X | X | 96 | 96 | 97 | 98 |
| R-507 | 97 | 98 | 98 | 97 | 97 | 96 | 95 | 97 | 97 | 96 | 96 | 80 | X | 96 | 96 | 98 | 98 |
| AC9000 | 98 | 98 | 98 | 98 | 98 | 98 | 98 | 97 | 97 | 96 | 97 | 98 | 98 | X | X | 98 | 98 |
| AZ20 | 98 | 98 | 98 | 98 | 98 | 97 | 98 | 97 | 97 | 96 | 96 | 96 | 96 | 96 | X | X | 98 |
| R-500 | 98 | 93 | 98 | 98 | 98 | 97 | 98 | 97 | 97 | 97 | 97 | 97 | 98 | 97 | 97 | X | X |
| R-502 | 97 | 98 | 96 | 97 | 97 | 98 | 98 | 95 | 95 | 95 | 94 | 97 | 98 | 97 | 97 | 98 | X |
| | R-12 | R-22 | R-125 | R-134a | R-115 | R-152a | R-143a | Propane | R-124 | R-32 | | | | | | | |
| R-401A | | 53 | | | | 13 | | | | | | | | | | | |
| MP66 | | 61 | | | | 11 | | | | | | | | | | | |
| R-402A | | 38 | 60 | | | | | | | | | | | | | | |
| R-402B | | 60 | 38 | | | | | | | | | | | | | | |
| R-404 | | | 44 | 4 | | | 52 | | | | | | | | | | |
| R-507 | | | 50 | | | | 50 | | | | | | | | | | |
| AC9000 | | | 25 | 52 | | | | | | 23 | | | | | | | |
| AZ20 | | | 50 | | | | | | | 50 | | | | | | | |
| R-500 | 73.8 | | | | | 26.2 | | | | | | | | | | | |
| R-502 | | 48.8 | | | 51.2 | | | | | | | | | | | | |

*Table 7-3. A 98% equivalent chart for several refrigerants*

# References

[1] SAE J1661, 1993, *Procedure for Retrofitting CFC-12 (R-12) Mobile Air Conditioning Systems to HFC-134a (R-134a)*.

[2] *Handling and Reuse of Refrigerants in the United States*, ARI, 1994.

[3] Kenneth W. Manz, "Survey of Residual Refrigerant 12 in Vehicle Air Conditioners Retrofit To R-134a." International CFC and Halo Alternatives Conference in Washington, D.C., September, 1992.

[4] SAE J2197, *HFC-134a Service Hose Fittings for Automotive Air Conditioning Service Equipment*, June, 1992.

[5] U.S. Patent 5,063,749, Robinair Division, SPX Corporation, 1991.

[6] U.S. Patent 5,181,391, Robinair Division, SPX Corporation, 1993.

[7] U.S. Patent 5,286,647, Robinair Division, SPX Corporation, 1994.

[8] U.S. Patent Pending, Robinair Division, SPX Corporation.

[9] U.S. Patent 5,158,747, Robinair Division, SPX Corporation, 1992.

Chapter Eight

# Deciding How to Process Refrigerant for Reuse

There are currently four choices[1] available for processing refrigerants before reuse:

1. Put refrigerant back into the system without recycling it.
2. Recycle refrigerant and put it back into the system it was removed from, or back into a system with the same owner or into a similar system if the used refrigerant remains in the contractor's custody at all times from recovery through charging into the new owner's system.
3. Recycle the refrigerant and test to verify conformance to ARI Standard 700 prior to reuse in a different owner's equipment (provided the refrigerant remains in the contractor's custody and control at all times from recovery through recycling to reuse).
4. Recover refrigerant for reclamation.

Each option is associated with a basic investment in equipment to recover the refrigerant; however, equipment and practices vary for each of the options. In general, both the refrigerant quality and cost of processing increase from option 1 to option 4.

## Selling Used Refrigerant

As stated in Chapter 7, mixed refrigerant is a big issue when selling used refrigerants, especially when using "arm's length" transactions such as between contractors. Other contaminants such as moisture, non-condensables, and acids affect the purity of the refrigerant. No other refrigerant conservation issue has produced such controversy as the sale of used refrigerant. System manufacturers and reclaimers have argued that selling used refrigerant should be banned unless it has been reclaimed to purity levels specified in ARI 700.

Some appliance manufacturers and chemical producers feel that reclaimed refrigerant is still not as clean as new refrigerant and should not be used in some systems. Motor vehicle manufacturers endorse selling recycled refrigerant of slightly lesser purity, so long as the refrigerant comes from another vehicle. Contractors and recycling equipment manufacturers claim that limited recycling and in-system clean-up procedures have been used successfully for many years.

Concerning tolerance design, Clausing wrote, "To exaggerate only slightly, the design engineers sit on one side of the conference table and chant 'tighter tolerances, tighter tolerances,' while the production engineers sit on the opposite side of the table and chant 'looser tolerances, looser tolerances.' The production engineers know that tighter tolerances cost more during production, and the design engineers know that tighter tolerances will provide operation that is closer to the ideal function of the product and will reduce the quality loss in the field. Usually the increase in unit manufacturing cost to provide tighter tolerances is well know, but the reduction in the expected quality loss in the field is not known; hence the irrational concurrence meetings."[2]

This analogy from a non-refrigerant-related field strikes close to home when applied to the issue of the quality of refrigerant offered for resale.

Pamphlet IRG-1, *Handling and Reuse of Refrigerants in the United States*,[1] covers decisions on reuse of recovered and recycled refrigerants. At an industry meeting hosted by the Air Conditioning Contractors of America (ACCA), recommended changes to IRG-1 were developed that would allow contractors recycling options for their own customers. Revision of the guideline to accommodate new recycling options is still expected in 1994. One contractor group, the Air Conditioning Contractors of America (ACCA), objected to the proposed requirement that "used refrigerant be analyzed and found to meet ARI 700 levels before being resold," because it would restrict contractors' options. The pamphlet, endorsed by nine organizations, is a guideline or industry-recommended practice, but it is not a legal document with any type of enforcement. As of this writing, EPA is studying whether any regulations are required after May, 1995. Industry guidelines, along with the contractor's reputation and the system manufacturer's warranty policy can be expected to lead to responsible service procedures.

At an historic meeting on October 6, 1994, participants agreed to supersede IRG-1 with IRG-2. The main change was to add option 3 (see page 105) and to revise the flowchart (see Figure 8-1). The ACCA board has endorsed IRG-2. Subsequently, EPA indicated plans to limit selling refrigerant to options 3 and 4 in upcoming regulations.

# Regulations and Recommendations

For motor vehicle air conditioners (MVACs), EPA Section 609 regulations require that used refrigerant removed from MVACs be recycled by equipment certified to meet SAE standards or be reclaimed before it is charged back into MVACs.

For all other refrigeration systems, the assumption is made that EPA Section 608 regulations will not address refrigerant purity after the *Sunset Clause* (prohibiting sale of refrigerant to a new owner unless it has been reclaimed) ends in May 1995. **Always follow all federal, state, and local regulations.**

Refrigeration system and component manufacturers publish warranty policies and recommended service procedures. Most manufacturers require charging new or reclaimed refrigerant during the warranty period, except when charging refrigerant back into the system from which it was removed for non-severe service conditions. **Always follow manufacturer recommendations.**

When industries address environmental problems to minimize societal impact and maximize environmental protection, they can avoid the necessity for some regulation. Pamphlet IRG-2[4] provides guidance on selecting one of the four options listed at the beginning of the chapter. Trade associations, recovery-recycling equipment manufacturers, reclaimers, and consulting engineers are sources of additional information. **Always follow industry-recommended practices.**

## Condition of Refrigerant

The reason the system is being serviced will help the service person determine the condition of the recovered refrigerant. Refrigerant component failures that do not affect refrigerant purity could allow the recovered refrigerant to be returned to the system without recycling. Compressor failures, especially motor burnouts, will weigh heavily toward a decision to recycle or reclaim the refrigerant.

If the system history is not known, the recovered refrigerant should, at a minimum, be recycled before it is put back into the system. If the system has had previous compressor problems that required replacing the compressor, then the service person should determine if it was a compressor burnout. The cleanliness of the used refrigerant from a previous burnout will depend on how well the system was cleaned up when the compressor was replaced.

Because systems with compressor burnouts are going to have some contaminants in the residual oil left in the system, the refrigerant should be as clean as possible so it does not add to the contamination that is already in the system. It is, therefore, a good practice to use recycled, new, or reclaimed refrigerant when a compressor burnout has occurred.

Burnout conditions may be determined by the characteristic burnt smell of oil or by checking the oil with an acid test kit. Tests such as vapor pressure measurement or boiling point determination can screen for mixed refrigerant conditions. Field instruments for recycling (see Chapter 7) can provide a more accurate determination.

## Recycling

If the refrigerant needs to be recycled before it can be put into a system, it should be cleaned to the recycled refrigerant contaminant levels in Table 8-1.

Recycling equipment should be certified to ARI Standard 740 and capable of consistently cleaning refrigerant to the contaminant levels in Table 8-1. The contaminated refrigerant sample used in ARI Standard 740 is representative of a highly contaminated system, so recycling equipment that can clean refrigerant in this test to Table 8-1 levels has acceptable cleaning capabilities.

It is important to periodically analyze recycled refrigerant to check the equipment's cleaning capability over time and ensure that its cleaning performance has not diminished. In addition, properly maintain recycling equipment to ensure maximum performance. Filter systems need to be changed or cleaned regularly, depending on the contamination of the refrigerant. Oil must be drained regularly from oil separators.

To this end, two equipment features are recommended:

- A filter-change indicator, which uses an indicator light or automatic shutdown to indicate that filters need changing (based on monitoring refrigerant moisture levels or monitoring the amount of refrigerant processed)
- An automatic clearing sequence, such as those described in Chapter 7

When all of the other factors are favorable, recycling should be feasible for the contractor and agreeable to the system owner. Contractors should consider time and effort required to recycle refrigerant versus recovering and reclaiming. In many cases, the quantity of refrigerant will impact this decision.

| Contaminants | Low Pressure Systems | R-12 Refrig. Systems | All Other Systems |
|---|---|---|---|
| Acid Content (by wt.) | 1.0 PPM | 1.0 PPM | 1.0 PPM |
| Moisture (by wt.) | 20 PPM | 10 PPM | 20 PPM |
| Non Condensable Gas (by Vol.) | N/A | 2.0% | 2.0% |
| High Boiling Residues (by Vol.) | 1.0% | 0.02% | 0.02% |
| Chlorides by Silver Nitrate Test | no turbidity | no turbidity | no turbidity |
| Particulates | visually clean | visually clean | visually clean |
| Other Refrigerants | 2.0% | 2.0% | 2.0% |

Note: To ensure that the recycling equipment maintains its demonstrated capability to achieve the above levels, it must be operated and maintained per the equipment manufacturer's recommendations.

Table 8-1. Maximum contaminant levels of recycled refrigerants in the same owner's equipment[4]

Whether charging with used refrigerant or new-reclaimed refrigerant, the system must be properly cleaned and evacuated prior to putting refrigerant back into the system. Manufacturers' recommended service procedures should be followed to ensure proper removal of contaminants. At a minimum, replace all driers. For systems with compressor burnouts, add suction line filters to assist in removing acids that remain in the system oil.

## Used Refrigerant Guidelines

The flowchart in Figure 8-1 is from the pamphlet *Handling and Reuse of Refrigerants in the United States*.[4] The following paragraphs discuss each of the options on the right side of the flowchart.

### Recovering for Reuse (Option 1)

Refrigerant can be recovered and returned to the same system if:

- the reason for current service is not for, nor does the service history indicate, a compressor motor burnout or other severe service conditions;
- the condition of the refrigerant as indicated by a system moisture indicator and vapor pressure check or other screening method is satisfactory.

Recovery equipment used for the purpose of returning the refrigerant to the same system should process the refrigerant through an oil separator. Separating oil will greatly reduce the contaminants returned to the system; minimize cross contamination from one system to another; and prolong the life of the recovery equipment.

Some of the advantages of option 1 include the following:

- Recovery-only equipment less expensive than recycling
- Requires fewest storage containers
- Lowest refrigerant-handling costs
- Takes less time than recycling
- Minimizes mixing due to consolidation

Disadvantages of option 1 include the following:

- Higher contaminant levels
- Options limited to same system
- Additional system clean-up required

Option 1 should not be used when transferring refrigerant from one system to another unless it has been analyzed and found to meet Table 8-1 contaminant levels.

# The Challenge of Recycling Refrigerants

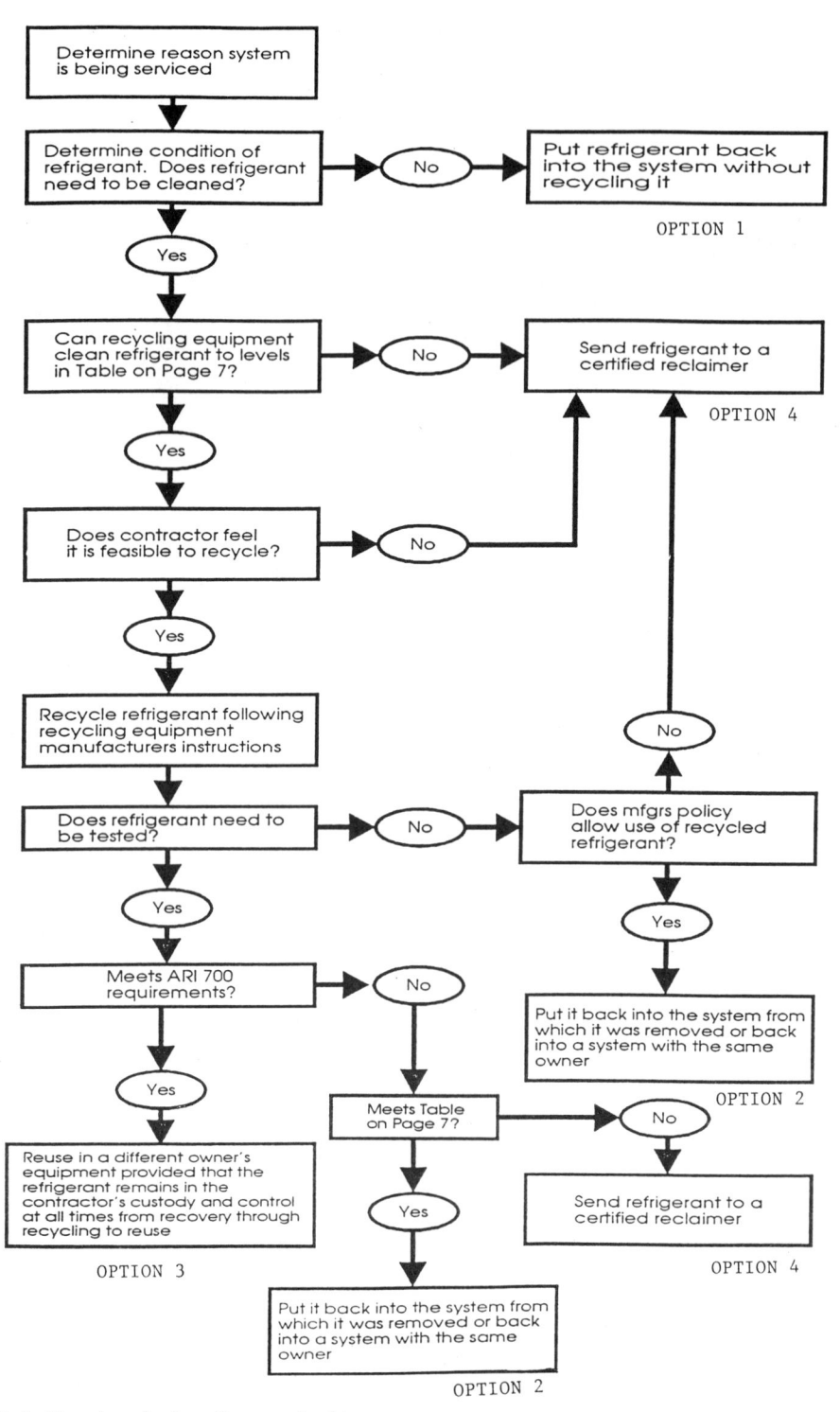

Figure 8-1. Flowchart for handling used refrigerants

## Recycling for Reuse (Option 2)

Refrigerants that are recovered from refrigeration and air conditioning systems may, at the system owner's option, be reused in that same owner's systems, as long as the contaminant levels of those refrigerants do not exceed the maximum levels shown in Table 8-1. Recycling equipment may be capable of cleaning refrigerant from a compressor burnout if a mixed refrigerant condition does not exist. However, follow manufacturers' policies, particularly for warranty repairs.

The contaminants of most concern in refrigerant to be recycled are "other refrigerants and organic impurities," as mixed refrigerants cannot be separated. In addition to screening tests, such as vapor pressure or boiling point measurement, suspected mixed refrigerants should be analyzed prior to recycling using field instruments or laboratory analysis (see Chapter 7).

If the refrigerant is to be removed from a system, recycled, and returned to the system, the recovery tanks must be kept clean so that recycled refrigerant does not become contaminated again when it enters the recovery tank. Follow the recycling equipment manufacturer instructions to clean recovery tanks. Purge any non-condensables in the recovery tank according to EPA rules and recycling equipment manufacturer instructions to prevent them from entering the system when the recycled refrigerant is added.

Refrigerant cleaning capability of recycling equipment is tested using the ARI Standard 740 test method, when operated according to the recycling equipment manufacturer instructions. Change filters, drain oil, and clear the equipment of previous refrigerant as recommended to achieve the same low contaminant levels achieved under the test conditions.

Some of the advantages of option 2 include the following:

- Low contaminant levels
- Reduced system clean-up
- Lower handling cost than reclaim
- Contractor and owner keep refrigerant
- Lower mixing potential

Disadvantages of option 2 include the following:

- High equipment investment
- Most time required
- Options limited to same owner

Option 2 should not be used when selling or using refrigerant in a different owner's equipment.

## Recycling and Verifying by Test (Option 3)

Refrigerants that are recovered for recycling at the contractor's shop may be reused in a different owner's equipment if:

- the recycled refrigerant batch has been tested and verified to meet ARI Standard 700 purity levels;
- the same contractor retains custody of the used refrigerant at all times from recovery through reuse.

The contractor will need to identify batches and retain records to identify which batch he charged out of. The recycling equipment will be similar to option 2. Preferably, this equipment would be certified to meet ARI 700 levels (at minimum Table 8-1) under ARI 740 test conditions. The recycling equipment will need to be maintained; analyzing each batch will serve to guide proper equipment operation.

Some of the advantages of option 3 include the following:

- Lowest contaminant levels
- Lowest system clean-up requirements
- Contractor keeps refrigerant
- Moderate handling costs
- Batching sensible for smaller systems (R-22)

Disadvantages of option 3 include the following:

- Highest equipment investment
- Most time required
- Cost of refrigerant analysis for small batches
- Keeping tanks clean

Option 3 allows the enterprising contractor to clean his own refrigerant while protecting the equipment manufacturer and owner.

## Recovering for Reclaim (Option 4)

When recovering for reclaim, concern for contaminant levels is minimized and recovery speed should be optimized. Liquid transfer such as the push-pull or liquid pump methods (see Chapter 3) work well for systems in which liquid refrigerant can be accessed. Keep different refrigerants in separate containers and label the containers.

Contractors should only send used refrigerants to reclaimers who participate in the ARI Reclaimed Refrigerant Certification Program (or equivalent outside the U.S.), to ensure that the reclaimer can properly process the refrigerant. *Under no circumstances should used refrigerant be sent to another contractor who recycles his own refrigerant and occasionally sends refrigerant for laboratory analysis at an owner's request.*

Reclaimed refrigerants offer the lowest contaminant levels and highest quality assurance. When the used refrigerant is sent to the reclaimer in the owner's container, the container will be properly cleaned. Since consolidation can cause cross contamination, refrigerant should be analyzed prior to consolidation using field instruments (see Chapter 7). Reclaimers should offer to pay more for tested refrigerant to encourage proper refrigerant-handling practices while protecting the used-refrigerant supply.

Some of the advantages of option 4 include the following:

- Lowest equipment costs
- Convenient for contractor
- Lowest contaminant levels
- More storage containers
- Lowest system clean-up requirements
- Lowest time required

Disadvantages of option 4 include the following:

- Highest handling costs
- Higher mixing potential
- Highest overall costs

Option 4 may be used without restriction for the same system, a different system with the same owner, new and warranty installations, and different owners.

## Summary

The options for processing refrigerants are recovery for reuse (option 1); recovery for recycle (option 2); recycle and verify by test (option 3); and recovery for reclaim (option 4). Follow government regulations, equipment manufacturer policies, and industry-recommended practices.

Factors to consider when deciding which option to use include reason for current service, condition of the refrigerant, recovery-recycling equipment cleaning capability, owner's preference, and clean-up of the system.

Option 1 is recommended only for the same system; the recovery equipment should include an oil separator. Option 2 is recommended for systems with the same owner. The recycling equipment should be capable of cleaning to Table 8-1 contaminant levels. Option 3 can be used for any of the contractor's customers but not for selling used refrigerant to another contractor or to the general public. Option 4 can be used in all circumstances, the same as new refrigerant. Large contractors may wish to employ all options. The rule of thumb is to clean the refrigerant as best you can.

# References

[1] *IRG-1 Handling and Reuse of Refrigerants in the United States*, ARI, 1994.

[2] *Total Quality Development, A Step-By-Step Guide to World-Class Concurrent Engineering*, Don Clausing, 1994.

[3] ARI Standard 700-93, *For Specifications for Fluorocarbon Refrigerants*, Copyright 1993.

[4] *IRG-2 Handling and Reuse of Refrigerants in the United States*, ARI, 1994.

Chapter Nine

# Choosing a Recovery-Recycling Unit for Stationary Applications

Recovery and recycling equipment comes in a variety of sizes, shapes, functions, and price ranges. It is important to investigate candidate equipment to make sure that it best meets the service technician's needs. A general knowledge of refrigeration systems and components, along with information contained in this book, provide the background you'll need to conduct this research. Select equipment based on the following points:

- Review your service operation, including the number of refrigerants handled and number of service vehicles, and determine which of options 1, 2, 3, 4 (Chapter 8) or a combination make sense. The typical charge amounts in these systems and their locations also are strong considerations.
- Investigate different types of equipment to determine which best meets your requirements. Equipment ratings under the ARI Standard 740[1] test methods will help you make an objective determination.
- Review equipment features, such as automatic operation, portability, and equipment monitors, to ensure design performance under field conditions. Consider service and technical support.
- Select equipment that meets regulatory requirements and your needs.

## ARI Standard 740 Rating Parameters

In the ARI Standard 740[1] test method, contaminated refrigerant (the so-called "dirty cocktail") is processed to determine the contaminant levels of the recycled refrigerant. The recovery and recycling rates and other parameters are determined.

EPA-approved third-party certification programs, such as ARI's *Refrigerant Recovery-Recycling Equipment Certification* program, provide quality comparisons between equipment models. Participants' model ratings, as appropriate, are included in the *Directory of Certified Refrigerant Recovery-Recycling Equipment*, published twice a year by ARI. Some of the information located in the *Directory* is as follows:

- **Type of equipment** — Equipment may be rated *recovery*, *system-dependent recovery*, *recovery-recycle*, and *recycle*. System-dependent recovery equipment may require an operating system compressor to function.
- **Designated refrigerants** — Refrigerants are listed for which the machine was rated at the time of manufacture. As new refrigerants come out, contact the equipment manufacturer concerning processing of these refrigerants.
- **Liquid recovery rate** — The rate at which the equipment recovers liquid refrigerant feeding the unit under test conditions. Because equipment tested using a push-pull method (see Chapter 3) may provide a rating 20 times higher than the same equipment tested using a vapor pumpout method, footnotes are shown for push-pull measurements that state, "Denotes manufacturer-selected push-pull method of liquid-recovery rating, which may not be applicable in all field applications." You cannot reach conclusions when comparing liquid recovery rates when one unit is rated by push-pull and another is rated by vapor pumpout methods.
- **Vapor recovery rate** — The next edition of ARI 740 (1994) is expected to list the average rate of removing vapor refrigerant from a tank starting at 75°F (24°C) test conditions and ending just below 0 psig (1 atm). For R-22, this test will be repeated at 104°F (40°C).
- **Final recovery vacuum** — The final recovery vacuum measured one minute after stopping the recovery sequence is listed; it must meet EPA regulations.
- **Recycle rate** — For equipment that uses a separate recycling sequence, the recycle rate is the amount of refrigerant cleaned divided by the time it takes to clean it. For equipment that cleans while recovering, the recycle rate is a maximum rate based solely on the higher of the liquid or vapor recovery rates, by which the rated contaminant levels can be attained.
- **Trapped refrigerant** — The amount of refrigerant that is left in a recovery or recycling unit rated for more than one refrigerant. The unit is cleared per the manufacturer's instructions before the measurement is taken. This is the amount that would cross-contaminate the next batch of a different refrigerant.

## Equipment Requirements

While not listed in the directory, all listed equipment must meet the following requirements:

- Operating instructions, necessary maintenance procedures, and source information for replacement parts and repair must be provided.
- The equipment shall indicate, usually by means of a transducer, indicator, or flowmeter, when any filter-driers need replacement (see Chapter 5 for comparison of methods).
- If purging is performed, the equipment shall automatically purge non-condensables or alert the operator that purging is required.
- Refrigerant emitted to the atmosphere during air purge, refrigerant clearing, and draining oil shall not exceed 3% by weight.

- Refrigerant clearing procedures must be accomplished in less than 15 minutes and, except for a vacuum pump and manifold gauge set, use components furnished by the equipment manufacturer.
- Hoses used with the equipment must be tested and meet low-permeation rates.

While there are no current levels required in ARI 740, contaminant levels should meet those shown in Table 8-1 in order to engage in recycling for reuse. These include the following:

- Moisture level in recycled refrigerant should be less than 20 ppm by weight (10 ppm for R-12) at the time the filter-drier is changed.
- Acid level in recycled refrigerant should be less than 1 ppm by weight.
- High-boiling residues (oil) in recycled refrigerant should be less than 0.02% (1% for low-pressure refrigerants such as R-11) by volume.
- Non-condensable gases (air) should be less than 2% by volume (not applicable for low-pressure refrigerants).

The following contaminant levels are not listed in the directory but must be met in order to list the type of equipment as *recycle* or *recover-recycle*:

- The recycled refrigerant must be clear of particulates and solids.
- Chloride ions in the recycled refrigerant must be less than 3 ppm by weight.

Regarding recovery equipment, certification ratings for recovery rates are realistic and good for comparison purposes, with caution in considering liquid recovery ratings using the push-pull method. The trapped refrigerant rating is significant for comparing units that handle multiple refrigerants.

Concerning recycling equipment, the recycling rate and contaminant levels are most significant. Obtaining the levels[1] recommended in Table 8-1 will significantly help contractors and service technicians uniformly interpret the ratings.

# Recovery Equipment — Operational Requirements

Some of the operational requirements to consider when evaluating recovery equipment are as follows:

- **Portability** — Contractors clearly prefer hand-carried units weighing 40 lb (18 kg) or less, which they can carry up a ladder and onto a roof. Such small units cannot handle all jobs, however, as current compressor technology and weight limit their recovery rates to about 0.5 lb per minute. When connected in a push-pull manner, liquid recovery rates are much faster when liquid is accessible. Referring to phase 2 vapor removal (see Chapter 5), some push-pull units are so slow they are not practical and encourage terminating recovery operation before achieving EPA final vacuum requirements.

- **Recovery time** — Contractors obviously prefer fast recovery rates and short recovery time. When recovering for reclaim (option 4), fast recovery rates are second in importance to portability. When recovering for reuse (option 1), the need to separate oil should have preference over time. In considering recovery time over many jobs, consider phase 1 liquid removal and phase 2 vapor removal (see Figure 5-9). If the contractor services a significant portion of systems that do not have liquid accessible, the contractor should consider selecting equipment based on vapor pumpout capacity.
- **Oil separation** — Since used oil is a source of cross-contamination and accelerated storage container corrosion, it is best to keep oil out of tanks, except when the refrigerant is sent to a reclaimer where the tanks will be cleaned. Oil must be drained frequently for proper operation.
- **Tank-overfill protection** — It is important to select and properly use equipment with overfill protection, to protect the operator and the environment.
- **Safety testing** — While ARI Standard 740 does not address safety concerns, it is advisable to use equipment that is evaluated and listed under UL 1963 equipment safety standards.
- **Storage containers** — Use proper containers (see Chapter 3); dedicate and label containers for single refrigerants.
- **Durability** — Equipment must be durable, compatible with the refrigerants and lubricants, and able to operate in the intended ambient conditions it will encounter. It should also be tight and free of leaks.
- **Use with multiple refrigerants** — Recovery units should be easy to clear and have a minimal amount of trapped refrigerant after the clearing process. When comparing the amount of trapped refrigerant for candidate equipment, consider equipment size and recovery rates.

# Recycling Equipment — Operational Requirements

Some of the operational requirements to consider when evaluating recycling equipment are as follows:

- **Refrigerant cleaning capabilities** — The most important requirement is the ability to clean refrigerant to meet industry guidelines[1] (see Table 8-1). Certification ratings (ARI 740 latest version) will determine if the equipment meets all contaminant levels except 2% "other refrigerants or organic impurities." Mixed refrigerants should be considered separately. The equipment must be maintained according to the manufacturer's instructions to achieve similar field results.
- **Filter replacement** — Contractors desire recycling equipment that has a long interval between filter changes and reliably indicates when to replace the filter. System manufacturers are concerned that the filter-driers won't be changed, allowing contaminated refrigerant to be placed in systems. The replacement interval is determined by the drier shape, desiccant amount and type, and its placement in the recycling equipment. Ease of changing filters varies with equipment design. Using chemical salt moisture indicators (see Chapter 5) for

filter replacement indicators is a matter of concern for equipment manufacturers. More reliable methods (moisture transducer, in-situ mass flowmeter, and scale) are discussed in Chapter 5. The indicator should show when to replace the drier without operator interpretation *and* terminate equipment operation.
- **Recycle rate** — The speed at which the recycling equipment recycles refrigerant is important, as well as the portion of time the unit must be attended.
- **Air purge** — It is important to purge only when necessary to meet non-condensable contaminant levels and to minimize refrigerant loss during the purge. Purge indicators and purge efficiency were covered in Chapter 6.
- **Multiple-refrigerant screening and clearing** — Recycling equipment normally will be used with multiple refrigerants. Built-in monitors to screen for mixed refrigerant conditions and identify the refrigerant type would enhance recycling unit operation. The recycling unit should be cleared after every batch to minimize cross-contamination (see Chapter 7).
- **Draining oil** — Since recycling equipment may be used with large batches of used refrigerants, the equipment should have a sight glass or other monitor to show when to stop and drain the oil. It is important to recover as much refrigerant as possible before draining the oil.
- **Portability** — Since most recycling will be performed at the shop, recycling equipment on a hand-cart or four-wheel dolly is considered appropriate.

## Federal Equipment Requirements

Under EPA Section 608 regulations, owners must have certified their refrigerant recovery and recycling equipment by August 12, 1993. Those who have not done so are still encouraged to, even though that deadline has passed. The certification form is shown in Figure 9-1.

Figure 9-2 shows the entire text of Section 608. Subsection 82.158 discusses standards for recycling and recovery equipment. In summary, equipment must:

- meet final recovery vacuum levels under ARI 740 test conditions;
- be certified by an approved equipment testing organization (currently ARI or UL);
- meet minimum requirements for ARI certification under ARI 740-93 (Since EPA adopted the latest version of ARI 740, all equipment requirements under Standard 740 are required by federal regulations.);
- have refrigerant loss when purging non-condensables less than 5% by weight until May 14, 1995 and less than 3% thereafter under ARI 740 test conditions;
- bear a special EPA label if manufactured after November 17, 1993.

Part 82 Subpart F is also shown in Figure 9-2. The most pertinent sections are 82.154 *Prohibitions*, 82.156 *Required Practices*, 82.161 *Technician Certification*, and 82.166 *Reporting and Recordkeeping Requirements*. For more information, contact the EPA Hotline at 800-296-1996.

The Challenge of Recycling Refrigerants

OMB # 2060-0256
Expiration Date: 5/96

## THE UNITED STATES ENVIRONMENTAL PROTECTION AGENCY (EPA) REFRIGERANT RECOVERY OR RECYCLING DEVICE ACQUISITION CERTIFICATION FORM

EPA regulations require establishments that service or dispose of refrigeration or air conditioning equipment to certify by August 12, 1993 that they have acquired recovery or recycling devices that meet EPA standards for such devices. To certify that you have acquired equipment, please complete this form according to the instructions and **mail it to the appropriate EPA Regional Office.** BOTH THE INSTRUCTIONS AND MAILING ADDRESSES CAN BE FOUND ON THE REVERSE SIDE OF THIS FORM.

### PART 1: ESTABLISHMENT INFORMATION

Name of Establishment

Street

(Area Code) Telephone Number

City    State    Zip Code

Number of Service Vehicles Based at Establishment

County

### PART 2: REGULATORY CLASSIFICATION

Identify the type of work performed by the establishment. **Check all boxes that apply.**

- ☐ Type A - Service small appliances
- ☐ Type B - Service refrigeration or air conditioning equipment other than small appliances
- ☐ Type C - Dispose of small appliances
- ☐ Type D - Dispose of refrigeration or air conditioning equipment other than small appliances

### PART 3: DEVICE IDENTIFICATION

| | Name of Device(s) Manufacturer | Model Number | Year | Serial Number (if any) | Check Box if Self-Contained |
|---|---|---|---|---|---|
| 1. | | | | | ☐ |
| 2. | | | | | ☐ |
| 3. | | | | | ☐ |
| 4. | | | | | ☐ |
| 5. | | | | | ☐ |
| 6. | | | | | ☐ |
| 7. | | | | | ☐ |

### PART 4: CERTIFICATION SIGNATURE

I certify that the establishment in Part 1 has acquired the refrigerant recovery or recycling device(s) listed in Part 2, that the establishment is complying with Section 608 regulations, and that the information given is true and correct.

Signature of Owner/Responsible Officer    Date    Name (Please Print)    Title

reporting burden for this collection of information is estimated to vary from 20 minutes to 60 minutes per response with an average of 40 minutes per response, including time for reviewing instructions, "ing existing data sources, gathering and maintaining the data needed, and completing the collection of information. Send comments regarding ONLY the burden estimates or any other aspects of this collection rmation, including suggestions for reducing this burden to Chief, Information Policy Branch, EPA, 401 M St., S.W. (PM-223Y); Washington, DC 20460, and to the Office of Information and Regulatory Affairs, of Management and Budget, Washington, DC 20503, marked "Attention: Desk Officer of EPA" DO NOT SEND THIS FORM TO THE ABOVE ADDRESSES. ONLY SEND COMMENTS TO THESE ! SSES

*Figure 9-1. Stationary appliance recycling equipment certification form (Source: EPA)*

## Choosing a Recovery-Recycling Unit for Stationary Applications

*Instructions*

Part 1: Please provide the name, address, and telephone number of the establishment where the refrigerant recovery or recycling device(s) is (are) located. Please complete one form for each location. State the number of vehicles based at this location that are used to transport technicians and equipment to and from service sites.

Part 2: Check the appropriate boxes for the type of work performed by technicians who are employees of the establishment. The term "small appliance" refers to any of the following products that are fully manufactured, charged, and hermetically sealed in a factory with five pounds or less of refrigerant: refrigerators and freezers designed for home use, room air conditioners (including window air conditioners and packaged terminal air conditioners), packaged terminal heat pumps, dehumidifiers, under-the-counter ice makers, vending machines, and drinking water coolers.

Part 3: For each recovery or recycling device acquired, please list the name of the manufacturer of the device, and (if applicable) its model number and serial number.

If more than 7 devices have been acquired, please fill out an additional form and attach it to this one. Recovery devices that are self-contained should be listed first and should be identified by checking the box in the last column on the right. Self-contained recovery equipment means refrigerant recovery or recycling equipment that is capable of removing the refrigerant from an appliance without the assistance of components contained in the appliance. On the other hand, system-dependent recovery equipment means refrigerant recovery equipment that requires the assistance of components contained in an appliance to remove the refrigerant from the appliance.

If the establishment has been listed as Type B and/or Type D in Part 2, then the first device listed in Part 3 must be a self-contained device and identified as such by checking the box in the last column on the right.

If any of the devices are homemade, they should be identified by writing "homemade" in the column provided for listing the name of the device manufacturer. Type A or Type B establishments can use homemade devices manufactured before November 15, 1993. Type C or Type D establishments can use homemade devices manufactured anytime. If, however, a Type C or Type D establishment is using homemade equipment manufactured after November 15, 1993, then it must not use these devices for service jobs.

Part 4: This form must be signed by either the owner of the establishment or another responsible officer. The person who signs is certifying that the establishment has acquired the equipment, that the establishment is complying with Section 608 regulations, and that the information provided is true and correct.

*EPA Regional Offices*

Send your form to the EPA office listed under the state or territory in which the establishment is located.

Connecticut, Maine, Massachusetts, New Hampshire, Rhode Island, Vermont

    CAA 608 Enforcement Contact: EPA Region I, Mail Code APC, JFK Federal Building, One Congress Street, Boston, MA 02203

New York, New Jersey, Puerto Rico, Virgin Islands

    CAA 608 Enforcement Contact: EPA Region II, Jacob K. Javits Federal Building, Room 5000, 26 Federal Plaza, New York, NY 10278

Delaware, District of Columbia, Maryland, Pennsylvania, Virginia, West Virginia

    CAA 608 Enforcement Contact: EPA Region III, Mail Code 3AT21, 841 Chestnut Building, Philadelphia, PA 19107

Alabama, Florida, Georgia, Kentucky, Mississippi, North Carolina, South Carolina, Tennessee

    CAA 608 Enforcement Contact: EPA Region IV, Mail Code APT-AE, 345 Courtland Street, NE, Atlanta, GA 30365

Illinois, Indiana, Michigan, Minnesota, Ohio, Wisconsin

    CAA 608 Enforcement Contact: EPA Region V, Mail Code AT18J, 77 W. Jackson Blvd., Chicago, IL 60604

Arkansas, Louisiana, New Mexico, Oklahoma, Texas

    CAA 608 Enforcement Contact: EPA Region VI, Mail Code 6T-EC, First Interstate Tower at Fountain Place, 1445 Ross Ave., Suite 1200, Dallas TX 75202

Iowa, Kansas, Missouri, Nebraska

    CAA 608 Enforcement Contact: EPA Region VII, Mail Code ARTX/ARBR, 726 Minnesota Ave., Kansas City, KS 66101

Colorado, Montana, North Dakota, South Dakota, Utah, Wyoming

    CAA 608 Enforcement Contact: EPA Region VIII, Mail Code 8AT-AP, 999 18th Street, Suite 500, Denver, CO 80202

American Samoa, Arizona, California, Guam, Hawaii, Nevada

    CAA 608 Enforcement Contact: EPA Region IX, Mail Code A-3, 75 Hawthorne Street, San Francisco, CA 94105

Alaska, Idaho, Oregon, Washington

    CAA 608 Enforcement Contact: EPA Region X, Mail Code AT-082, 1200 Sixth Ave., Seattle, WA 98101

**Subpart F—Recycling and Emissions Reduction**

**§ 82.150  Purpose and scope.**

(a) The purpose of this subpart is to reduce emissions of class I and class II refrigerants to the lowest achievable level during the service, maintenance, repair, and disposal of appliances in accordance with section 608 of the Clean Air Act.

(b) This subpart applies to any person servicing, maintaining, or repairing appliances except for motor vehicle air conditioners. This subpart also applies to persons disposing of appliances, including motor vehicle air conditioners. In addition, this subpart applies to refrigerant reclaimers, appliance owners, and manufacturers of appliances and recycling and recovery equipment.

**§ 82.152  Definitions.**

(a) *Appliance* means any device which contains and uses a class I or class II substance as a refrigerant and which is used for household or commercial purposes, including any air conditioner, refrigerator, chiller, or freezer.

(b) *Approved equipment testing organization* means any organization which has applied for and received approval from the Administrator pursuant to § 82.160.

(c) *Certified refrigerant recovery or recycling equipment* means equipment certified by an approved equipment testing organization to meet the standards in § 82.158 (b) or (d), equipment certified pursuant to § 82.36(a), or equipment manufactured before November 15, 1993, that meets the standards in § 82.158 (c), (e), or (g).

(d) *Commercial refrigeration* means, for the purposes of § 82.156(i), the refrigeration appliances utilized in the retail food and cold storage warehouse sectors. Retail food includes the refrigeration equipment found in supermarkets, convenience stores, restaurants and other food service establishments. Cold storage includes the equipment used to store meat, produce, dairy products, and other perishable goods. All of the equipment contains large refrigerant charges, typically over 75 pounds.

(e) *Disposal* means the process leading to and including:

(1) The discharge, deposit, dumping or placing of any discarded appliance into or on any land or water;

(2) The disassembly of any appliance for discharge, deposit, dumping or placing of its discarded component parts into or on any land or water; or

(3) The disassembly of any appliance for reuse of its component parts.

(f) *High-pressure appliance* means an appliance that uses a refrigerant with a boiling point between $-50$ and $10$ degrees Centigrade at atmospheric pressure (29.9 inches of mercury). This definition includes but is not limited to appliances using refrigerants $-12$, $-22$, $-114$, $-500$, or $-502$.

(g) *Industrial process refrigeration* means, for the purposes of § 82.156(i), complex customized appliances used in the chemical, pharmaceutical, petrochemical and manufacturing industries. This sector also includes industrial ice machines and ice rinks.

(h) *Low-loss fitting* means any device that is intended to establish a connection between hoses, appliances, or recovery or recycling machines and that is designed to close automatically or to be closed manually when disconnected, minimizing the release of refrigerant from hoses, appliances, and recovery or recycling machines.

(i) *Low-pressure appliance* means an appliance that uses a refrigerant with a boiling point above 10 degrees Centigrade at atmospheric pressure (29.9 inches of mercury). This definition includes but is not limited to equipment utilizing refrigerants $-11$, $-113$, and $-123$.

(j) *Major maintenance, service, or repair* means any maintenance, service, or repair involving the removal of any or all of the following appliance components: Compressor, condenser, evaporator, or auxiliary heat exchanger coil.

(k) *Motor vehicle air conditioner (MVAC)* means any appliance that is a motor vehicle air conditioner as defined in 40 CFR part 82, subpart B.

Figure 9-2. Federal EPA Section 608 regulations (Source: EPA)

(l) *MVAC-like appliance* means mechanical vapor compression, open-drive compressor appliances used to cool the driver's or passenger's compartment of an non-road motor vehicle. This includes the air-conditioning equipment found on agricultural or construction vehicles. This definition is not intended to cover appliances using HCFC–22 refrigerant.

(m) *Normally containing* a quantity of refrigerant means containing the quantity of refrigerant within the appliance or appliance component when the appliance is operating with a full charge of refrigerant.

(n) *Opening* an appliance means any service, maintenance, or repair on an appliance that could be reasonably expected to release refrigerant from the appliance to the atmosphere unless the refrigerant were previously recovered from the appliance.

(o) *Person* means any individual or legal entity, including an individual, corporation, partnership, association, state, municipality, political subdivision of a state, Indian tribe, and any agency, department, or instrumentality of the United States, and any officer, agent, or employee thereof.

(p) *Process stub* means a length of tubing that provides access to the refrigerant inside a small appliance or room air conditioner and that can be resealed at the conclusion of repair or service.

(q) *Reclaim* refrigerant means to reprocess refrigerant to at least the purity specified in the ARI Standard 700–1988, Specifications for Fluorocarbon Refrigerants (appendix A to 40 CFR part 82, subpart F) and to verify this purity using the analytical methodology prescribed in the ARI Standard 700–1988. In general, reclamation involves the use of processes or procedures available only at a reprocessing or manufacturing facility.

(r) *Recover* refrigerant means to remove refrigerant in any condition from an appliance without necessarily testing or processing it in any way.

(s) *Recovery efficiency* means the percentage of refrigerant in an appliance that is recovered by a piece of recycling or recovery equipment.

(t) *Recycle* refrigerant means to extract refrigerant from an appliance and clean refrigerant for reuse without meeting all of the requirements for reclamation. In general, recycled refrigerant is refrigerant that is cleaned using oil separation and single or multiple passes through devices, such as replaceable core filter-driers, which reduce moisture, acidity, and particulate matter. These procedures are usually implemented at the field job site.

(u) *Self-contained recovery equipment* means refrigerant recovery or recycling equipment that is capable of removing the refrigerant from an appliance without the assistance of components contained in the appliance.

(v) *Small appliance* means any of the following products that are fully manufactured, charged, and hermetically sealed in a factory with five (5) pounds or less of refrigerant: refrigerators and freezers designed for home use, room air conditioners (including window air conditioners and packaged terminal air conditioners), packaged terminal heat pumps, dehumidifiers, under-the-counter ice makers, vending machines, and drinking water coolers.

(w) *System-dependent recovery equipment* means refrigerant recovery equipment that requires the assistance of components contained in an appliance to remove the refrigerant from the appliance.

(x) *Technician* means any person who performs maintenance, service, or repair that could reasonably be expected to release class I or class II substances from appliances into the atmosphere, including but not limited to installers, contractor employees, in-house service personnel, and in some cases, owners. Technician also means any person disposing of appliances except for small appliances.

(y) *Very high-pressure appliance* means an appliance that uses a refrigerant with a boiling point below $-50$ degrees Centigrade at atmospheric pressure (29.9 inches of mercury). This definition includes but is not limited to equipment utilizing refrigerants $-13$ and $-503$.

### § 82.154 Prohibitions.

(a) Effective June 14, 1993, no person maintaining, servicing, repairing, or disposing of appliances may knowingly vent or otherwise release into the environment any class I or class II substance used as refrigerant in such equipment. De minimis releases associated with good faith attempts to recycle or recover refrigerants are not subject to this prohibition. Releases shall be considered de minimis if they occur when:

(1) The required practices set forth in § 82.156 are observed and recovery or recycling machines that meet the requirements set forth in § 82.158 are used; or

(2) The requirements set forth in 40 CFR part 82, subpart B are observed. The knowing release of refrigerant subsequent to its recovery from an appliance shall be considered a violation of this prohibition.

(b) Effective July 13, 1993 no person may open appliances except MVACs for maintenance, service, or repair, and no person may dispose of appliances except for small appliances, MVACs, and MVAC-like appliances:

(1) Without observing the required practices set forth in § 82.156; and

(2) Without using equipment that is certified for that type of appliance pursuant to § 82.158.

(c) Effective November 15, 1993, no person may manufacture or import recycling or recovery equipment for use during the maintenance, service, or repair of appliances except MVACs, and no person may manufacture or import recycling or recovery equipment for use during the disposal of appliances except small appliances, MVACs, and MVAC-like appliances, unless the equipment is certified pursuant to § 82.158 (b), (d), or (f), as applicable.

(d) Effective June 14, 1993, no person shall alter the design of certified refrigerant recycling or recovery equipment in a way that would affect the equipment's ability to meet the certification standards set forth in § 82.158 without resubmitting the altered design for certification testing. Until it is tested and shown to meet the certification standards set forth in § 82.158, equipment so altered will be considered uncertified for the purposes of § 82.158.

(e) Effective August 12, 1993, no person may open appliances except MVACs for maintenance, service, or repair, and no person may dispose of appliances except for small appliances, MVACs, and MVAC-like appliances, unless such person has certified to the Administrator pursuant to § 82.162 that such person has acquired certified recovery or recycling equipment and is complying with the applicable requirements of this subpart.

(f) Effective August 12, 1993, no person may recover refrigerant from small appliances, MVACs, and MVAC-like appliances for purposes of disposal of these appliances unless such person has certified to the Administrator pursuant to § 82.162 that such person has acquired recovery equipment that meets the standards set forth in § 82.158 (l) and/or (m), as applicable, and that such person is complying with the applicable requirements of this subpart.

(g) Effective August 12, 1993 until November 13, 1995, no person may sell or offer for sale for use as a refrigerant any class I or class II substance consisting wholly or in part of used refrigerant unless the class I or class II substance has been reclaimed as defined at § 82.152(q).

(h) Effective August 12, 1993 until November 13, 1995, no person may sell or offer for sale for use as a refrigerant any class I or class II substance consisting wholly or in part of used refrigerant unless the refrigerant has been reclaimed by a person who has been certified as a reclaimer pursuant to § 82.164.

(i) Effective August 12, 1993, no person reclaiming refrigerant may release more than 1.5% of the refrigerant received by them.

(j) Effective November 15, 1993, no person may sell or distribute, or offer for sale or distribution, any appliances, except small appliances, unless such equipment is equipped with a servicing aperture to facilitate the removal of refrigerant at servicing and disposal.

(k) Effective November 15, 1993, no person may sell or distribute, or offer for

Figure 9-2. Federal EPA Section 608 regulations — continued

sale or distribution any small appliance unless such equipment is equipped with a process stub to facilitate the removal of refrigerant at servicing and disposal.

(l) Effective November 14, 1994 no person may open an appliance except for an MVAC and no person may dispose of an appliance except for a small appliance, MVAC, or MVAC-like appliance, unless such person has been certified as a technician for that type of appliance pursuant to § 82.161.

(m) No technician training or testing program may issue certificates pursuant to § 82.161 unless the program complies with all of the standards of § 82.161 and appendix D, and has been granted approval.

(n) Effective November 14, 1994 no person may sell or distribute, or offer for sale or distribution, any class I or class II substance for use as a refrigerant to any person unless:

(1) The buyer has been certified as a Type I, Type II, Type III, or Universal technician pursuant to § 82.161;

(2) The buyer has been certified pursuant to 40 CFR part 82, subpart B;

(3) The refrigerant is sold only for eventual resale to certified technicians or to appliance manufacturers (e.g., sold by a manufacturer to a wholesaler, sold by a technician to a reclaimer);

(4) The refrigerant is sold to an appliance manufacturer;

(5) The refrigerant is contained in an appliance; or

(6) the refrigerant is charged into an appliance by a certified technician during maintenance, service, or repair.

(o) It is a violation of this subpart to accept a signed statement pursuant to § 82.156(f)(2) if the person knew or had reason to know that such a signed statement is false.

### § 82.156  Required practices.

(a) Effective July 13, 1993, all persons opening appliances except for MVACs for maintenance, service, or repair must evacuate the refrigerant in either the entire unit or the part to be serviced (if the latter can be isolated) to a system receiver or a recovery or recycling machine certified pursuant to § 82.158. All persons disposing of appliances except for small appliances, MVACs, and MVAC-like appliances must evacuate the refrigerant in the entire unit to a recovery or recycling machine certified pursuant to § 82.158.

(1) Persons opening appliances except for small appliances, MVACs, and MVAC-like appliances for maintenance, service, or repair must evacuate to the levels in Table 1 before opening the appliance, unless

(i) Evacuation of the appliance to the atmosphere is not to be performed after completion of the maintenance, service, or repair, and the maintenance, service, or repair is not major as defined at § 82.152(j); or

(ii) Due to leaks in the appliance, evacuation to the levels in Table 1 is not attainable, or would substantially contaminate the refrigerant being recovered. In any of these cases, the requirements of § 82.156(a)(2) must be followed.

(2)(i) If evacuation of the appliance to the atmosphere is not to be performed after completion of the maintenance, service, or repair, and if the maintenance, service, or repair is not major as defined at § 82.152(j), the appliance must:

(A) Be evacuated to a pressure no higher than 0 psig before it is opened if it is a high- or very high-pressure appliance; or

(B) Be pressurized to 0 psig before it is opened if it is a low-pressure appliance, without using methods, e.g., nitrogen, that require subsequent purging.

(ii) If, due to leaks in the appliance, evacuation to the levels in Table 1 is not attainable, or would substantially contaminate the refrigerant being recovered, persons opening the appliance must:

(A) Isolate leaking from non-leaking components wherever possible;

(B) Evacuate non-leaking components to be opened to the levels specified in Table 1; and

(C) Evacuate leaking components to be opened to the lowest level that can be attained without substantially contaminating the refrigerant. In no case shall this level exceed 0 psig.

(3) Persons disposing of appliances except for small appliances, MVACs, and MVAC-like appliances, must evacuate to the levels in Table 1.

TABLE 1.—REQUIRED LEVELS OF EVACUATION FOR APPLIANCES
[Except for small appliances, MVACs, and MVAC-like appliances]

| Type of appliance | Inches of Hg vacuum (relative to standard atmospheric pressure of 29.9 inches Hg) ||
| --- | --- | --- |
| | Using recovery or recycling equipment manufactured or imported before Nov. 15, 1993 | Using recovery or recycling equipment manufactured or imported on or after Nov. 15, 1993 |
| HCFC-22 appliance, or isolated component of such appliance, normally containing less than 200 pounds of refrigerant. | 0 | 0. |
| HCFC-22 appliance, or isolated component of such appliance, normally containing less than 200 pounds of refrigerant. | 0 | 0. |
| HCFC-22 appliance, or isolated component of such appliance, normally containing 200 pounds or more of refrigerant. | 4 | 10. |
| Other high-pressure appliance, or isolated component of such appliance, normally containing less than 200 pounds of refrigerant. | 4 | 10. |
| Other high-pressure appliance, or isolated component of such appliance, normally containing 200 pounds or more of refrigerant. | 4 | 15. |
| Very high-pressure appliance | 0 | 0. |
| Low-pressure appliance | 25 | 25 mm Hg absolute. |

(4) Persons opening small appliances for maintenance, service, or repair must:

(i) When using recycling and recovery equipment manufactured before November 15, 1993, recover 80% of the refrigerant in the small appliance; or

(ii) When using recycling or recovery equipment manufactured on or after November 15, 1993, recover 90% of the refrigerant in the appliance when the compressor in the appliance is operating, or 80% of the refrigerant in the appliance when the compressor in the appliance is not operating; or

(iii) Evacuate the small appliance to four inches of mercury vacuum.

(5) Persons opening MVAC-like appliances for maintenance, service, or repair may do so only while properly using, as defined at § 82.32(e), recycling or recovery equipment certified pursuant to § 82.158 (f) or (g), as applicable.

(b) Effective July 13, 1993, all persons opening appliances except for small appliances and MVACs for maintenance, service, or repair and all persons disposing of appliances except for small appliances must have at least one piece of certified, self-contained recovery equipment available at their place of business.

(c) System-dependent equipment shall not be used with appliances normally containing more than 15 pounds of refrigerant.

(d) All recovery or recycling equipment shall be used in accordance with the manufacturer's directions unless such directions conflict with the requirements of this subpart.

(e) Refrigerant may be returned to the appliance from which it is recovered or to another appliance owned by the same person without being recycled or reclaimed, unless the appliance is an MVAC-like appliance.

(f) Effective July 13, 1993, persons who take the final step in the disposal process (including but not limited to scrap recyclers and landfill operators) of a small appliance, room air conditioning, MVACs, or MVAC-like appliances must either:

(1) Recover any remaining refrigerant from the appliance in accordance with paragraph (g) or (h) of this section, as applicable; or

(2) Verify that the refrigerant has been evacuated from the appliance or shipment of appliances previously. Such verification must include a signed statement from the person from whom the appliance or shipment of appliances is obtained that all refrigerant that had not leaked previously has been recovered from the appliance or shipment of appliances in accordance with paragraph (g) or (h) of this section, as applicable. This statement must include the name and address of the person who recovered the refrigerant and the date the refrigerant was recovered or a contract that refrigerant will be removed prior to delivery.

*Figure 9-2. Federal EPA Section 608 regulations — continued*

(3) Persons complying with paragraph (f)(2) of this section must notify suppliers of appliances that refrigerant must be properly removed before delivery of the items to the facility. The form of this notification may be warning signs, letters to suppliers, or other equivalent means.

(g) All persons recovering refrigerant from MVACs and MVAC-like appliances for purposes of disposal of these appliances must reduce the system pressure to or below 102 mm of mercury vacuum, using equipment that meets the standards set forth in § 82.158(l).

(h) All persons recovering the refrigerant from small appliances for purposes of disposal of these appliances must either:

(1) Recover 90% of the refrigerant in the appliance when the compressor in the appliance is operating, or 80% of the refrigerant in the appliance when the compressor in the appliance is not operating; or

(2) Evacuate the small appliance to four inches of mercury vacuum.

(i) (1) Owners of commercial refrigeration and industrial process refrigeration equipment must have all leaks repaired if the equipment is leaking at a rate such that the loss of refrigerant will exceed 35 percent of the total charge during a 12 month period, except as described in paragraph (i)(3) of this section.

(2) Owners of appliances normally containing more than 50 pounds of refrigerant and not covered by paragraph (i)(1) of this section must have all leaks repaired if the appliance is leaking at a rate such that the loss of refrigerant will exceed 15 % of the total charge during a 12-month period, except as described in paragraph (i)(3) of this section.

(3) Owners are not required to repair the leaks defined in paragraphs (i)(1) and (2) of this section if, within 30 days, they develop a one-year retrofit or retirement plan for the leaking equipment. This plan (or a legible copy) must be kept at the site of the equipment. The original must be made available for EPA inspection on request. The plan must be dated and all work under the plan must be completed within one year of plan's date.

(4) Owners must repair leaks pursuant to paragraphs (i)(1) and (2) of this section within 30 days of discovery or within 30 days of when the leak(s) should have been discovered, if the owners intentionally shielded themselves from information which would have revealed a leak.

(Approved by the Office of Management and Budget under the control number 2060–0256)

## § 82.158 Standards for recycling and recovery equipment.

(a) Effective November 15, 1993, all manufacturers and importers of recycling and recovery equipment intended for use during the maintenance, service, or repair of appliances except MVACs and MVAC-like appliances or during the disposal of appliances except small appliances, MVACs, and MVAC-like appliances, shall have had such equipment certified by an approved equipment testing organization to meet the applicable requirements in paragraph (b) or (d) of this section. All manufacturers and importers of recycling and recovery equipment intended for use during the maintenance, service, or repair of MVAC-like appliances shall have had such equipment certified pursuant to § 82.36(a).

(b) Equipment manufactured or imported on or after November 15, 1993 for use during the maintenance, service, or repair of appliances except small appliances, MVACs, and MVAC-like appliances or during the disposal of appliances except small appliances, MVACs, and MVAC-like appliances must be certified by an approved equipment testing organization to meet the following requirements:

(1) In order to be certified, the equipment must be capable of achieving the level of evacuation specified in Table 2 of this section under the conditions of the ARI Standard 740–1993, Performance of Refrigerant Recovery, Recycling and/or Reclaim Equipment (ARI 740–1993) (Appendix B):

TABLE 2.—LEVELS OF EVACUATION WHICH MUST BE ACHIEVED BY RECOVERY OR RECYCLING EQUIPMENT INTENDED FOR USE WITH APPLIANCES [1]

[Manufactured on or after November 15, 1993]

| Type of appliance with which recovery or recycling machine is intended to be used | Inches of Hg vacuum |
|---|---|
| HCFC-22 appliances, or isolated component of such appliances, normally containing less than 200 pounds of refrigerant | 0 |
| HCFC-22 appliances, or isolated component of such appliances, normally containing 200 pounds or more of refrigerant | 10 |
| Very high-pressure appliances | 0 |
| Other high-pressure appliances, or isolated component of such appliances, normally containing less than 200 pounds of refrigerant | 10 |
| Other high-pressure appliances, or isolated component of such appliances, normally containing 200 pounds or more of refrigerant | 15 |
| Low-pressure appliances | [2] 25 |

[1] Except for small appliances, MVACs, and MVAC-like appliances.
[2] mm Hg absolute.

The vacuums specified in inches of Hg vacuum must be achieved relative to an atmospheric pressure of 29.9 inches of Hg absolute.

(2) Recovery or recycling equipment whose recovery efficiency cannot be tested according to the procedures in ARI 740–1993 may be certified if an approved third-party testing organization adopts and performs a test that demonstrates, to the satisfaction of the Administrator, that the recovery efficiency of that equipment is equal to or better than that of equipment that:

(i) Is intended for use with the same type of appliance; and
(ii) Achieves the level of evacuation in Table 2.

(3) The equipment must meet the minimum requirements for ARI certification under ARI 740–1993.

(4) If the equipment is equipped with a noncondensables purge device:

(i) The equipment must not release more than five percent of the quantity of refrigerant being recycled through noncondensables purging under the conditions of ARI 740–1993; and

(ii) Effective May 14, 1995, the equipment must not release more than three percent of the quantity of refrigerant being recycled through noncondensables purging under the conditions of ARI 740–1993.

(5) The equipment must be equipped with low-loss fittings on all hoses.

(6) The equipment must have its liquid recovery rate and its vapor recovery rate measured under the conditions of ARI 740–1993.

(c) Equipment manufactured or imported before November 15, 1993 for use during the maintenance, service, or repair of appliances except small appliances, MVACs, and MVAC-like appliances or during the disposal of appliances except small appliances, MVACs, and MVAC-like appliances will be considered certified if it is capable of achieving the level of evacuation specified in Table 3 of this section when tested using a properly calibrated pressure gauge:

TABLE 3.—LEVELS OF EVACUATION WHICH MUST BE ACHIEVED BY RECOVERY OR RECYCLING MACHINES INTENDED FOR USE WITH APPLIANCES [1]

[Manufactured before November 15, 1993]

| Type of air-conditioning or refrigeration equipment with which recovery or recycling machine is intended to be used | Inches of vacuum (relative to standard atmospheric pressure of 29.9 inches Hg) |
|---|---|
| HCFC-22 equipment, or isolated component of such equipment, normally containing less than 200 pounds of refrigerant | 0 |
| HCFC-22 equipment, or isolated component of such equipment, normally containing 200 pounds or more of refrigerant | 4 |
| Very high-pressure equipment | 0 |
| Other high-pressure equipment, or isolated component of such equipment, normally containing less than 200 pounds of refrigerant | 4 |
| Other high-pressure equipment, or isolated component of such equipment, normally containing 200 pounds or more of refrigerant | 4 |
| Low-pressure equipment | 25 |

[1] Except for small appliances, MVACs, and MVAC-like appliances.

*Figure 9-2. Federal EPA Section 608 regulations — continued*

(d) Equipment manufactured or imported on or after November 15, 1993 for use during the maintenance, service, or repair of small appliances must be certified by an approved equipment testing organization to be capable of either:

(1) Recovering 90% of the refrigerant in the test stand when the compressor of the test stand is operating and 80% of the refrigerant when the compressor of the test stand is not operating when used in accordance with the manufacturer's instructions under the conditions of appendix C, Method for Testing Recovery Devices for Use with Small Appliances; or

(2) Achieving a four-inch vacuum under the conditions of appendix B, ARI 740–1993.

(e) Equipment manufactured or imported before November 15, 1993 for use with small appliances will be considered certified if it is capable of either:

(1) Recovering 80% of the refrigerant in the system, whether or not the compressor of the test stand is operating, when used in accordance with the manufacturer's instructions under the conditions of appendix C, Method for Testing Recovery Devices for Use with Small Appliances; or

(2) Achieving a four-inch vacuum when tested using a properly calibrated pressure gauge.

(f) Equipment manufactured or imported on or after November 15, 1993 for use during the maintenance, service, or repair of MVAC-like appliances must be certified in accordance with § 82.36(a).

(g) Equipment manufactured or imported before November 15, 1993 for use during the maintenance, service, or repair of MVAC-like appliances must be capable of reducing the system pressure to 102 mm of mercury vacuum under the conditions of the SAE Standard, SAE J1990 (appendix A to 40 CFR part 82, subpart B).

(h) Manufacturers and importers of equipment certified under paragraphs (b) and (d) of this section must place a label on each piece of equipment stating the following:

THIS EQUIPMENT HAS BEEN CERTIFIED BY [APPROVED EQUIPMENT TESTING ORGANIZATION] TO MEET EPA's MINIMUM REQUIREMENTS FOR RECYCLING OR RECOVERY EQUIPMENT INTENDED FOR USE WITH [APPROPRIATE CATEGORY OF APPLIANCE].

The label shall also show the date of manufacture and the serial number (if applicable) of the equipment. The label shall be affixed in a readily visible or accessible location, be made of a material expected to last the lifetime of the equipment, present required information in a manner so that it is likely to remain legible for the lifetime of the equipment, and be affixed in such a manner that it cannot be removed from the equipment without damage to the label.

(i) The Administrator will maintain a list of equipment certified pursuant to paragraphs (b), (d), and (f) of this section by manufacturer and model. Persons interested in obtaining a copy of the list should send written inquiries to the address in § 82.160(a).

(j) Manufacturers or importers of recycling or recovery equipment intended for use during the maintenance, service, or repair of appliances except MVACs or MVAC-like appliances or during the disposal of appliances except small appliances, MVACs, and MVAC-like appliances must periodically have approved equipment testing organizations conduct either:

(1) Retests of certified recycling or recovery equipment; or

(2) Inspections of recycling or recovery equipment at manufacturing facilities to ensure that each equipment model line that has been certified under this section continues to meet the certification criteria.

Such retests or inspections must be conducted at least once every three years after the equipment is first certified.

(k) An equipment model line that has been certified under this section may have its certification revoked if it is subsequently determined to fail to meet the certification criteria. In such cases, the Administrator or her or his designated representative shall give notice to the manufacturer or importer setting forth the basis for her or his determination.

(l) Equipment used to evacuate refrigerant from MVACs and MVAC-like appliances before they are disposed of must be capable of reducing the system pressure to 102 mm of mercury vacuum under the conditions of the SAE Standard, SAE J1990 (appendix A to 40 CFR part 82, subpart B).

(m) Equipment used to evacuate refrigerant from small appliances before they are disposed of must be capable of either:

(1) Removing 90% of the refrigerant when the compressor of the small appliance is operating and 80% of the refrigerant when the compressor of the small appliance is not operating, when used in accordance with the manufacturer's instructions under the conditions of appendix C, Method for Testing Recovery Devices for Use With Small Appliances; or

(2) Evacuating the small appliance to four inches of vacuum when tested using a properly calibrated pressure gauge.

### § 82.160 Approved equipment testing organizations.

(a) Any equipment testing organization may apply for approval by the Administrator to certify equipment pursuant to the standards in § 82.158 and appendices B or C of this subpart. The application shall be sent to: Section 608 Recycling Program Manager, Stratospheric Protection Division, 6205–J, U.S. Environmental Protection Agency, 401 M Street, SW., Washington, DC 20460.

(b) Applications for approval must include written information verifying the following:

(1) The list of equipment present at the organization that will be used for equipment testing.

(2) Expertise in equipment testing and the technical experience of the organization's personnel.

(3) Thorough knowledge of the standards as they appear in § 82.158 and appendices B and/or C (as applicable) of this subpart.

(4) The organization must describe its program for verifying the performance of certified recycling and recovery equipment manufactured over the long term, specifying whether retests of equipment or inspections of equipment at manufacturing facilities will be used.

(5) The organization must have no conflict of interest and receive no direct or indirect financial benefit from the outcome of certification testing.

(6) The organization must agree to allow the Administrator access to records and personnel to verify the information contained in the application.

(c) Organizations may not certify equipment prior to receiving approval from EPA. If approval is denied under this section, the Administrator or her or his designated representative shall give written notice to the organization setting forth the basis for her or his determination.

(d) If at any time an approved testing organization is found to be conducting certification tests for the purposes of this subpart in a manner not consistent with the representations made in its application for approval under this section, the Administrator reserves the right to revoke approval. In such cases, the Administrator or her or his designated representative shall give notice to the organization setting forth the basis for her or his determination.

(e) Testing organizations seeking approval of an equipment certification program may also seek approval to certify equipment tested previously under the program. Interested organizations may submit to the Administrator at the address in § 82.160(a) verification that the program met all of the standards in § 82.160(b) and that equipment to be certified was tested to and met the applicable standards in § 82.158 (b) or (d). Upon EPA approval, the previously tested equipment may be certified without being retested (except insofar as such retesting is part of the testing organization's program for verifying the performance of equipment manufactured over the long term, pursuant to § 82.160(b)(4)).

(Approved by the Office of Management and Budget under the control number 2060–0256)

Figure 9-2. Federal EPA Section 608 regulations — continued

### § 82.161 Technician certification.

(a) Effective November 14, 1994, persons who maintain, service, or repair appliances, except MVACs, and persons who dispose of appliances, except for small appliances, room air conditioners, and MVACs, must be certified by an approved technician certification program as follows:

(1) Persons who maintain, service, or repair small appliances as defined in § 82.158(v) must be properly certified as Type I technicians.

(2) Persons who maintain, service, or repair high or very high-pressure appliances, except small appliances and MVACs, or dispose of high or very high-pressure appliances, except small appliances and MVACs, must be properly certified as Type II technicians.

(3) Persons who maintain, service, or repair low-pressure appliances or dispose of low-pressure appliances must be properly certified as Type III technicians.

(4) Persons who maintain, service, or repair low- and high-pressure equipment as described in § 82.161(a) (1), (2) and (3) must be properly certified as Universal technicians.

(5) Persons who maintain, service, or repair MVAC-like appliances must either be properly certified as Type II technicians or complete the training and certification test offered by a training and certification program approved under § 82.40.

(b) *Test Subject Material.* The Administrator shall maintain a bank of test questions divided into four groups, including a core group and three technical groups. The Administrator shall release this bank of questions only to approved technician certification programs. Tests for each type of certification shall include a minimum of 25 questions drawn from the core group and a minimum of 25 questions drawn from each relevant technical group. These questions shall address the subject areas listed in appendix D.

(c) *Program Approval.* Persons may seek approval of any technician certification program (program), in accordance with the provisions of this paragraph, by submitting to the Administrator at the address in § 82.160(a) verification that the program meets all of the standards listed in appendix D and the following standards:

(1) *Alternative Examinations.* Programs are encouraged to make provisions for non-English speaking technicians by providing tests in other languages or allowing the use of a translator when taking the test. If a translator is used, the certificate received must indicate that translator assistance was required. A test may be administered orally to any person who makes this request, in writing, to the program at least 30 days before the scheduled date for the examination. The letter must explain why the request is being made.

(2) *Recertification.* The Administrator reserves the right to specify the need for technician recertification at some future date, if necessary, by placing a notice in the **Federal Register**.

(3) *Proof of Certification.* Programs must issue individuals a wallet-sized card to be used as proof of certification, upon successful completion of the test. Programs must issue an identification card to technicians that receive a score of 70 percent or higher on the closed-book certification exam, within 30 days. Programs providing Type I certification using the mail-in format, must issue a permanent identification card to technicians that receive a score of 84 percent or higher on the certification exam, no later than 30 days after the program has received the exam and any additional required material. Each card must include, at minimum, the name of the certifying program, and the date the organization became a certifying program, the name of the person certified, the type of certification, a unique number for the certified person, and the following text:

[Name of person] has been certified as a [Type I, Type II, Type III, and/or Universal, as appropriate] technician as required by 40 CFR part 82, subpart F.

(4) The Administrator reserves the right to consider other factors deemed relevant to ensure the effectiveness of certification programs.

(d) If approval is denied under this section, the Administrator shall give written notice to the program setting forth the basis for her or his determination.

(e) If at any time an approved program violates any of the above requirements, the Administrator reserves the right to revoke approval. In such cases, the Administrator or her or his designated representative shall give notice to the organization setting forth the basis for her or his determination.

(f) Authorized representatives of the Administrator may require technicians to demonstrate on the business entity's premises their ability to perform proper procedures for recovering and/or recycling refrigerant. Failure to demonstrate or failure to properly use the equipment may result in revocation of the certificate. Failure to abide by any of the provisions of this subpart may also result in revocation or suspension of the certificate. If a technician's certificate is revoked, the technician would need to recertify before maintaining, servicing, repairing or disposing of any appliances.

(g) Persons seeking approval of a technician certification program may also seek approval for technician certifications granted previously under the program. Interested persons may submit to the Administrator at the address in § 82.160(a) verification that the program met all of the standards of § 82.161(c) and appendix D, or verification that the program met all of the standards of § 82.161(c) and appendix D, except for some elements of the test subject material, in which case the person must submit verification that supplementary information on that material will be provided pursuant to appendix D, section (j).

(Approved by the Office of Management and Budget under the control number 2060–0256)

### § 82.162 Certification by owners of recovery and recycling equipment.

(a) No later than August 12, 1993, or within 20 days of commencing business for those persons not in business at the time of promulgation, persons maintaining, servicing, or repairing

appliances except for MVACs, and persons disposing of appliances except for small appliances and MVACs, must certify to the Administrator that such person has acquired certified recovery or recycling equipment and is complying with the applicable requirements of this subpart. Such equipment may include system-dependent equipment but must include self-contained equipment, if the equipment is to be used in the maintenance, service, or repair of appliances except for small appliances. The owner or lessee of the recovery or recycling equipment may perform this certification for his or her employees. Certification shall take the form of a statement signed by the owner of the equipment or another responsible officer and setting forth:

(1) The name and address of the purchaser of the equipment, including the county name;

(2) The name and address of the establishment where each piece of equipment is or will be located;

(3) The number of service trucks (or other vehicles) used to transport technicians and equipment between the establishment and job sites and the field;

(4) The manufacturer name, the date of manufacture, and if applicable, the model and serial number of the equipment; and

(5) The certification must also include a statement that the equipment will be properly used in servicing or disposing of appliances and that the information given is true and correct. Owners or lessees of recycling or recovery equipment having their places of business in:

Connecticut
Maine
Massachusetts
New Hampshire
Rhode Island
Vermont

must send their certifications to:

CAA § 608 Enforcement Contact, EPA Region I, Mail Code APC, JFK Federal Building, One Congress Street, Boston, MA 02203.

Owners or lessees of recycling or recovery equipment having their places of business in:

New York
New Jersey
Puerto Rico
Virgin Islands

*Figure 9-2. Federal EPA Section 608 regulations — continued*

must send their certifications to:

CAA § 608 Enforcement Contact, EPA Region II, Jacob K. Javits Federal Building, 26 Federal Plaza, Room 5000, New York, NY 10278.

Owners or lessees of recycling or recovery equipment having their places of business in:

Delaware
District of Columbia
Maryland
Pennsylvania
Virginia
West Virginia

must send their certifications to:

CAA § 608 Enforcement Contact, EPA Region III, Mail Code 3AT21, 841 Chestnut Building, Philadelphia, PA 19107.

Owners or lessees of recycling or recovery equipment having their places of business in:

Alabama
Florida
Georgia
Kentucky
Mississippi
North Carolina
South Carolina
Tennessee

must send their certifications to:

CAA § 608 Enforcement Contact, EPA Region IV, 345 Courtland Street, NE., Mail Code APT-AE, Atlanta, GA 30365.

Owners or lessees of recycling or recovery equipment having their places of business in:

Illinois
Indiana
Michigan
Minnesota
Ohio
Wisconsin

must send their certifications to:

CAA § 608 Enforcement Contact, EPA Region V, Mail Code AT18J, 77 W. Jackson Blvd., Chicago, IL 60604-3507.

Owners or lessees of recycling or recovery equipment having their places of business in:

Arkansas
Louisiana
New Mexico
Oklahoma
Texas

must send their certifications to:

CAA § 608 Enforcement Contact, EPA Region VI, Mail Code 6T-EC, First Interstate Tower at Fountain Place, 1445 Ross Ave., Suite 1200, Dallas, TX 75202-2733.

Owners or lessees of recycling or recovery equipment having their places of business in:

Iowa
Kansas
Missouri
Nebraska

must send their certifications to:

CAA § 608 Enforcement Contact, EPA Region VII, Mail Code ARTX/ARBR, 726 Minnesota Ave., Kansas City, KS 66101.

Owners or lessees of recycling or recovery equipment having their places of business in:

Colorado
Montana
North Dakota
South Dakota
Utah
Wyoming

must send their certifications to:

CAA § 608 Enforcement Contact, EPA Region VIII, Mail Code 8AT-AP, 999 18th Street, Suite 500, Denver, CO 80202-2405.

Owners or lessees of recycling or recovery equipment having their places of business in:

American Samoa
Arizona
California
Guam
Hawaii
Nevada

must send their certifications to:

CAA § 608 Enforcement Contact, EPA Region IX, Mail Code A-3, 75 Hawthorne Street, San Francisco, CA 94105.

Owners or lessees of recycling or recovery equipment having their places of business in:

Alaska
Idaho
Oregon
Washington

must send their certifications to:

CAA § 608 Enforcement Contact, EPA Region X, Mail Code AT-082, 1200 Sixth Ave., Seattle, WA 98101.

(b) Certificates under paragraph (a) of this section are not transferable. In the event of a change of ownership of an entity that maintains, services, or repairs appliances except MVACs, or that disposes of appliances except small appliances, MVACs, and MVAC-like appliances, the new owner of the entity shall certify within 30 days of the change of ownership pursuant to paragraph (a) of this section.

(c) No later than August 12, 1993, persons recovering refrigerant from small appliances, MVACs, and MVAC-like appliances for purposes of disposal of these appliances must certify to the Administrator that such person has acquired recovery equipment that meets the standards set forth in § 82.158 (l) and/or (m), as applicable, and that such person is complying with the applicable requirements of this subpart. Such equipment may include system-dependent equipment but must include self-contained equipment, if the equipment is to be used in the disposal of appliances except for small appliances. The owner or lessee of the recovery or recycling equipment may perform this certification for his or her employees. Certification shall take the form of a statement signed by the owner of the equipment or another responsible officer and setting forth:

(1) The name and address of the purchaser of the equipment, including the county name;

(2) The name and address of the establishment where each piece of equipment is or will be located;

(3) The number of service trucks (or other vehicles) used to transport technicians and equipment between the establishment and job sites and the field;

(4) The manufacturer's name, the date of manufacture, and if applicable, the model and serial number of the equipment; and

(5) The certification must also include a statement that the equipment will be properly used in recovering refrigerant from appliances and that the information given is true and correct. The certification shall be sent to the appropriate address in paragraph (a).

(d) Failure to abide by any of the provisions of this subpart may result in revocation or suspension of certification under paragraph (a) or (c) of this section. In such cases, the Administrator or her or his designated representative shall give notice to the organization setting forth the basis for her or his determination.

(Approved by the Office of Management and Budget under the control number 2060-0256)

### § 82.164 Reclaimer certification.

Effective August 12, 1993, persons reclaiming used refrigerant for sale to a new owner must certify to the Administrator that such person will:

(a) Return refrigerant to at least the standard of purity set forth in ARI Standard 700-1988, Specifications for Fluorocarbon Refrigerants;

(b) Verify this purity using the methods set forth in ARI Standard 700-1988;

(c) Release no more than 1.5 percent of the refrigerant during the reclamation process, and

(d) Dispose of wastes from the reclamation process in accordance with all applicable laws and regulations. The data elements for certification are as follows:

(1) The name and address of the reclaimer;

(2) A list of equipment used to reprocess and to analyze the refrigerant; and

(3) The owner or a responsible officer of the reclaimer must sign the certification stating that the refrigerant will be returned to at least the standard of purity set forth in ARI Standard 700-1988, Specifications for Fluorocarbon Refrigerants, that the purity of the refrigerant will be verified using the methods set forth in ARI Standard 700-1988, that no more than 1.5 percent of the refrigerant will be released during the reclamation process, that wastes from the reclamation process will be properly disposed of, and that the information given is true and correct. The certification should be sent to the following address: Section 608 Recycling Program Manager, Stratospheric Protection Division,

*Figure 9-2. Federal EPA Section 608 regulations — continued*

(6205–J), U.S. Environmental Protection Agency, 401 M Street, SW., Washington, DC 20460.

(e) Certificates are not transferable. In the event of a change in ownership of an entity which reclaims refrigerant, the new owner of the entity shall certify within 30 days of the change of ownership pursuant to this section.

(f) Failure to abide by any of the provisions of this subpart may result in revocation or suspension of the certification of the reclaimer. In such cases, the Administrator or her or his designated representative shall give notice to the organization setting forth the basis for her or his determination.

(Approved by the Office of Management and Budget under the control number 2060–0256)

### § 82.166 Reporting and recordkeeping requirements.

(a) All persons who sell or distribute any class I or class II substance for use as a refrigerant must retain invoices that indicate the name of the purchaser, the date of sale, and the quantity of refrigerant purchased.

(b) Purchasers of any class I or class II refrigerants who employ technicians who recover refrigerants may provide evidence of each technician's certification to the wholesaler who sells them refrigerant; the wholesaler will then keep this information on file. In such cases, the purchaser must notify the wholesaler regarding any change in a technician's certification or employment status.

(c) Approved equipment testing organizations must maintain records of equipment testing and performance and a list of equipment that meets EPA requirements. A list of all certified equipment shall be submitted to EPA within 30 days of the organization's approval by EPA and annually at the end of each calendar year thereafter.

(d) Approved equipment testing organizations shall submit to EPA within 30 days of the certification of a new model line of recycling or recovery equipment the name of the manufacturer and the name and/or serial number of the model line.

(e) Approved equipment testing organizations shall notify EPA if retests of equipment or inspections of manufacturing facilities conducted pursuant to § 82.158(j) show that a previously certified model line fails to meet EPA requirements. Such notification must be received within thirty days of the retest or inspection.

(f) Programs certifying technicians must maintain records in accordance with section (g) of appendix D of this subpart.

(g) Reclaimers must maintain records of the names and addresses of persons sending them material for reclamation and the quantity of the material (the combined mass of refrigerant and contaminants) sent to them for reclamation. Such records shall be maintained on a transactional basis.

(h) Reclaimers must maintain records of the quantity of material sent to them for reclamation, the mass of refrigerant reclaimed, and the mass of waste products. Reclaimers must report this information to the Administrator annually within 30 days of the end of the calendar year.

(i) Persons disposing of small appliances, MVACs, and MVAC-like appliances must maintain copies of signed statements obtained pursuant to § 82.156(f)(2).

(j) Persons servicing appliances normally containing 50 or more pounds of refrigerant must provide the owner/operator of such appliances with an invoice or other documentation, which indicates the amount of refrigerant added to the appliance.

(k) Owners/operators of appliances normally containing 50 or more pounds of refrigerant must keep servicing records documenting the date and type of service, as well as the quantity of refrigerant added. The owner/operator must keep records of refrigerant purchased and added to such appliances in cases where owners add their own refrigerant. Such records should indicate the date(s) when refrigerant is added.

(l) Technicians certified under § 82.161 must keep a copy of their certificate at their place of business.

(m) All records required to be maintained pursuant to this section must be kept for a minimum of three years unless otherwise indicated. Entities that dispose of appliances must keep these records on-site.

(Approved by the Office of Management and Budget under the control number 2060–0256)

Figure 9-2. Federal EPA Section 608 regulations — concluded

## Summary

When selecting recovery-recycling equipment, review your service operation to define needs, investigate different equipment and features, and select equipment that meets regulatory requirements *and* your needs.

ARI Standard 740 (latest version) provides a standardized method of testing and rating recovery-recycling equipment. The performance of equipment manufactured by participants in third-party certification programs may be found in certification directories. Operational requirements for recovery and recycling equipment were discussed separately.

## Reference

[1] ARI, *Handling and Reuse of Refrigerants in the United States*, April 1994.

# Chapter Ten

# Motor Vehicle System Recycling and Retrofitting

In 1988 and 1989, the Society of Automotive Engineers (SAE) developed a refrigerant recycling equipment standard[1], a recycled refrigerant purity standard[2], and a recommended service procedure[3] to launch a recycling program for R-12.

An ad hoc group was assembled by the U.S. Environmental Protection Agency (EPA) consisting of vehicle manufacturers, component suppliers, chemical producers, recovery-recycling equipment manufacturers, environmentalists, and government regulators. This group outlined a research project to collect used refrigerant from a variety of automobile air conditioners in four regions of the United States.

Based on the results of this EPA-funded project, the group selected recycled refrigerant purity levels. A 15,000-mile (24,200-km) car with no known air conditioning problems was used as a baseline.

The SAE standards defined recycling equipment performance and safety requirements. These standards also outlined the equipment testing, certification, and labeling requirements to ensure proper operation under shop conditions. A schematic of the R-12 recycling program is shown in Figure 10-1.

The next major SAE task was to provide for servicing and recycling R-134a in 1990 and 1991. Standards similar to those for R-12 were written for recycling equipment,[4] recycled refrigerant purity,[5] and a service procedure.[6]

The SAE committee was very concerned about mixed refrigerant (see Chapter 7) due to early findings that polyalkaline glycol (PAG) lubricant was incompatible with the chlorine in R-12. They also had a desire to protect the used refrigerant supply from inadvertent mixing of refrigerant types.

To prevent mixing used R-12 with used R-134a, SAE developed non-threaded service ports on R-134a vehicles,[7] a unique R-134a service equipment fitting,[8] and a hose standard[9] to keep service equipment and containers separate for the two refrigerants. The approach shown in Figure 10-2 illustrates how R-12 equipment and procedures are used for R-12 and R-134a equipment and procedures are used for

The Challenge of Recycling Refrigerants

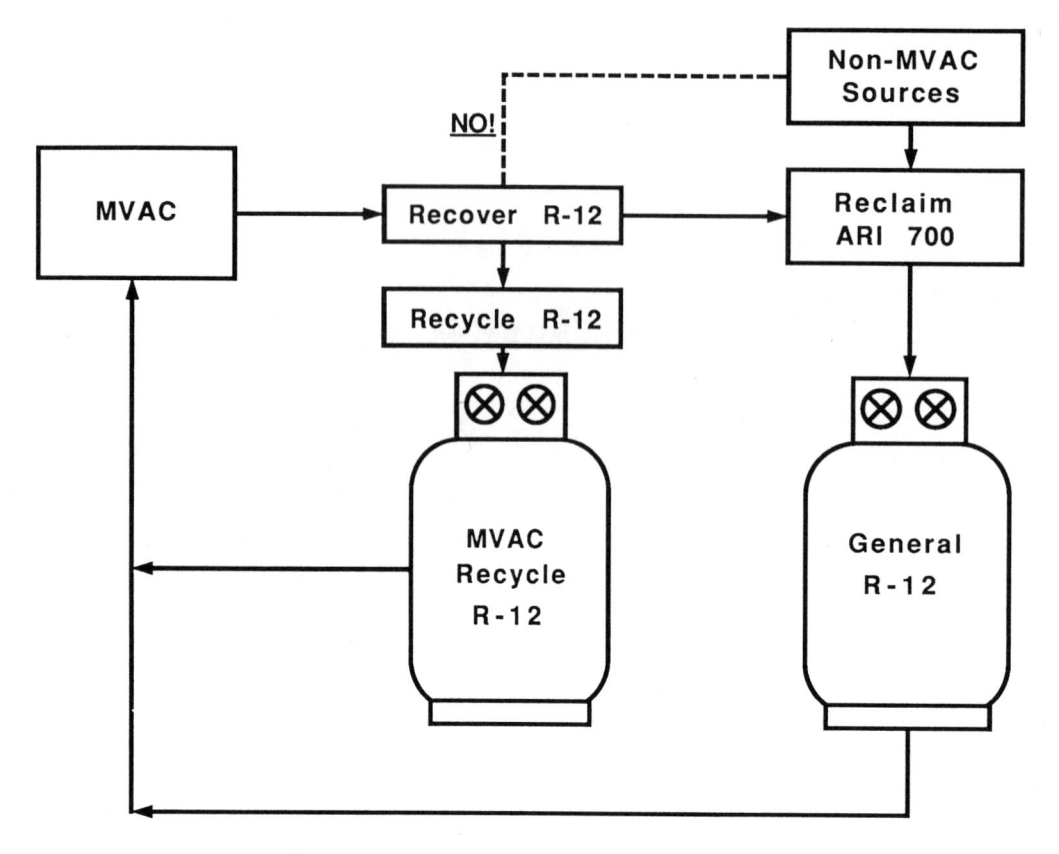

*Figure 10-1. Recycling R-12 (1987 to 1991)*

R-134a. Different service ports and service fittings help to keep this separation, Figure 10-3. Five years of field experience have shown this to work quite well.

# Retrofitting Vehicles from R-12 to R-134a

When the 1996 R-12 phaseout date was set, it became obvious that retrofitting would be required to adequately service the existing R-12 fleet. The SAE committee acted upon the recommendation of vehicle manufacturers to identify R-134a as the preferred retrofit refrigerant. The primary reasons for this are as follows:

- The extensive testing and development work with R-134a for new and retrofit vehicles
- Available service equipment for R-134a
- The ability to better manage the purity of used refrigerant supplies by staying with two service refrigerants

The approach taken was a logical extension of the R-12 and R-134a recycling programs, Figure 10-4. Since new R-134a vehicles have special fittings and a label

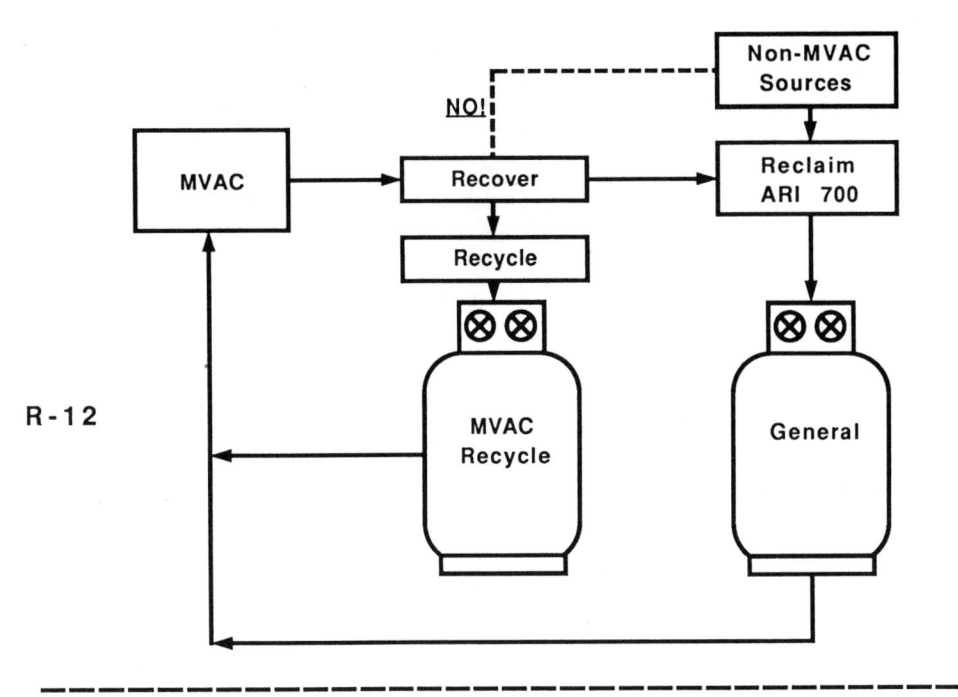

Motor Vehicle System Recycling and Retrofitting

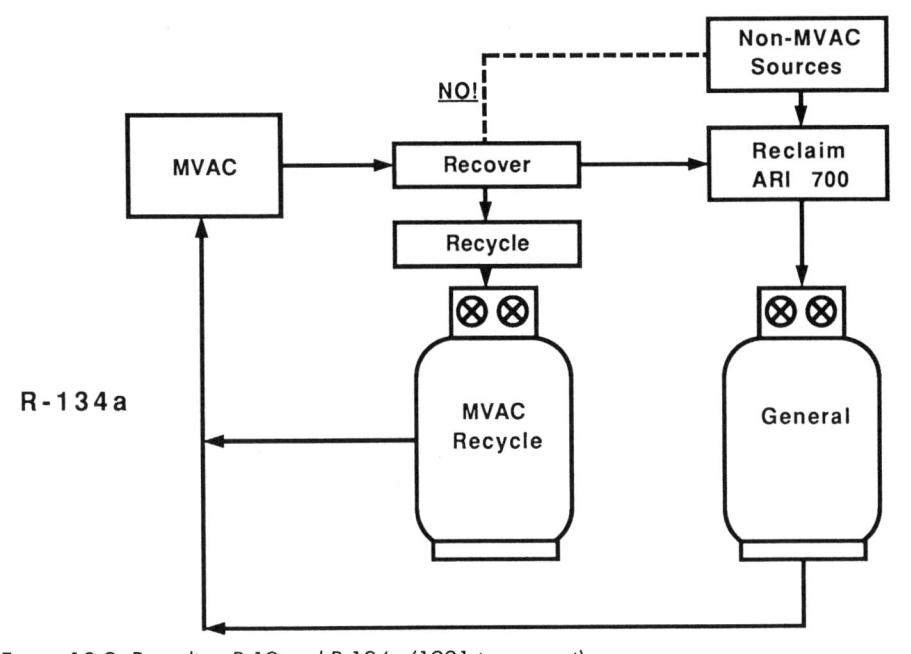

*Figure 10-2. Recycling R-12 and R-134a (1991 to present)*

*Figure 10-3. Automotive service fitting (Courtesy, Robinair Division, SPX Corporation)*

stating refrigerant and lubricant quantities, retrofit vehicles would also need the same fittings and a retrofit label.[10] Figuratively, R-12 service equipment still connects only to R-12 vehicles and R-134a service equipment connects to R-134a vehicles. Behind the "magic" retrofit curtain, the motor vehicle air conditioner changes from R-12 to R-134a.

The retrofit procedure[11] emphasizes following vehicle manufacturers' recommendations and pulling a deep vacuum to ensure full removal of the R-12. Flushing with R-12 to remove oil has been used successfully and is described as an option if full oil removal is required. On some vehicles, flushing connections are complicated and fully removing the R-12 used to flush may be difficult.

## Flushing Survey

SAE and industry conducted a survey[12] where refrigerant was sampled from retrofitted motor vehicle air conditioners (MVACs) on field tests. The samples were analyzed for residual R-12 levels, and the service procedures were reviewed, Table 10-1. The major considerations were:

- Was oil drained?
- Was desiccant replaced?
- Did they flush?
- Were other components replaced?
- How long and to what level did they pull a vacuum?

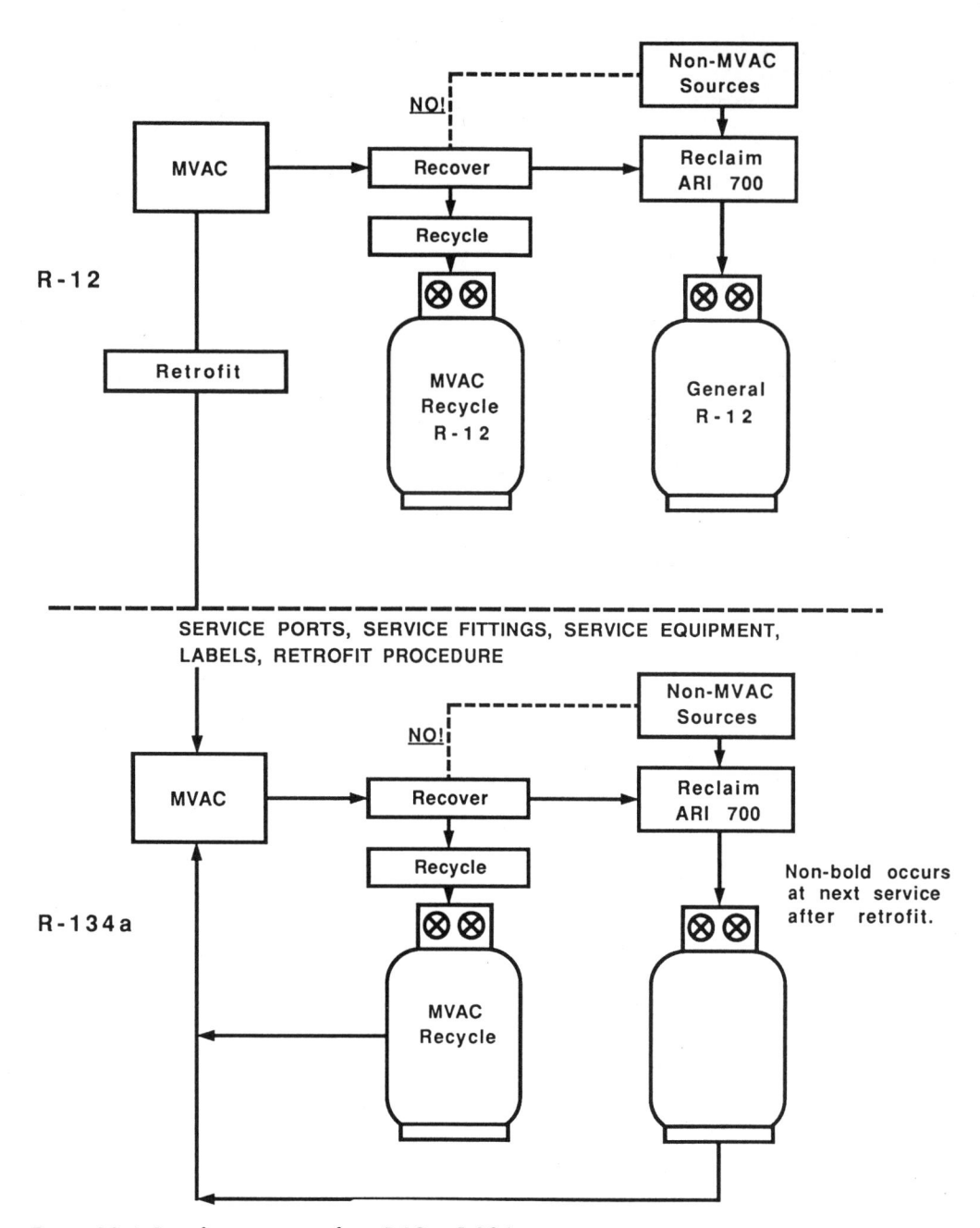

Figure 10-4. Retrofitting systems from R-12 to R-134a

The analysis determined the percentage by weight of R-12 in the R-134a. With two exceptions, those MVACs evacuated for at least 30 minutes contained levels less than 1% residual R-12. Those exceptions were large vehicles with dual air conditioning that had been flushed with R-12. Most likely, the R-12 used for flushing was not completely removed.

# The Challenge of Recycling Refrigerants

| SAMPLE | DRAIN OIL? | REPLACE DESICCANT | FLUSH ? | REPLACE COMPONENT | VACUUM TIME (MINUTES) | R-12 (percent) |
|---|---|---|---|---|---|---|
| B |  | X | X |  | 10 | .90 |
| D |  | X | X |  | 480 | .00 |
| E | X | X |  | X | 45 | .26 |
| F |  | X | X |  | 45 | .23 |
| H |  | X |  |  | 0 | 3.50 |
| J |  | ? |  | X | 60 | .82 |
| K |  | X |  |  | 45 | .49 |
| L | X |  |  |  | 30 | .32 |
| M | X | X |  |  | 15 | 1.04 |
| N |  | X | X |  | 45 | 3.14* |
| O |  | X | X |  | 45 | 1.70** |
| P |  | X | X |  | 45 | .91 |
| Q |  | X | X |  | 30 | .46 |
| R | X | X | X |  | 30 | .25 |
| S | X | X | X |  | 30 | .17 |
| T |  | X | X |  | 30 | .73 |
| U |  | X | X |  | 30 | .15 |
| V | X | X |  |  | 30 | .35 |
| W | X | X |  |  | 30 | .29 |
| X | X | X |  |  | 30 | .27 |
| Y | X | X |  |  | 30 | <.50 |
| Z | X | X |  |  | 30 | <.50 |
| AA |  | X |  |  | 6 | .75 |
| AB |  |  |  |  | 0 | 4.75 |

\* Vehicle had York Compressor with separate oil sump and very little oil was removed during flushing.

\*\* Vehicle was extended cab pick-up truck with front and rear air conditioning.

*Table 10-1. Outline of service procedures (Courtesy, SAE participants)*

# Combination R-12 and R-134a Recycling Units

As explained in Chapter 7, up to the current time the MVAC industry has chosen to use separate recycling equipment for R-12 and R-134a. SAE is in the process of writing and defining a combination R-12 and R-134a equipment standard. This equipment should be easy to connect, maintain separate fitting integrity, and meet the SAE cross-contamination requirements after switching refrigerants.

Motor Vehicle System Recycling and Retrofitting

Based on the experience gained in producing multiple-refrigerant recycling equipment for stationary equipment, some equipment manufacturers have pushed for combination units for MVACs. This is practical if proper safeguards are used to preserve the intent and positive experience received thus far with separate equipment.

Where common refrigerant circuits are used for both R-12 and R-134a, such equipment must:

- protect against inadvertent introduction of the wrong refrigerant to the wrong tank;
- clear the common circuit before switching from one refrigerant to the other.

One way to accomplish these objectives would be to provide separate inlet hoses for R-12, Figure 10-5, protected by valves 1 and 2 and separate outlet hoses to two separate storage containers protected by valves 3 and 4[13].

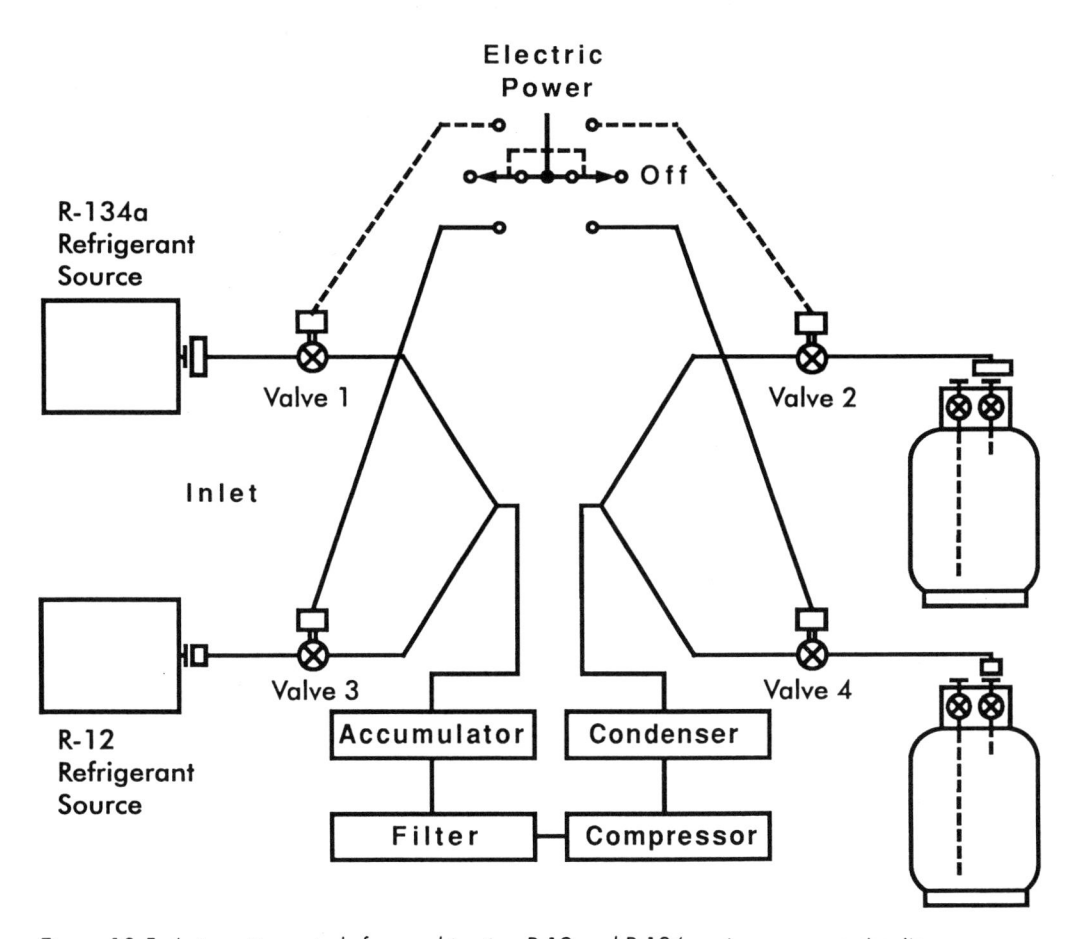

*Figure 10-5. Automatic controls for combination R-12 and R-134a using common circuits*

# The Challenge of Recycling Refrigerants

An R-12/R-134a selector switch would ensure that valves 1, 2, 3, and 4 work together to provide the proper connections. Connecting the wrong hose to the wrong vehicle or tank would be difficult because of the fittings. If an R-12 vehicle were connected and the switch selector was on R-134a, nothing would happen. Clearing methods have been thoroughly covered in Chapters 3 and 7. Air purge methods discussed in Chapter 6 must also be used.

Another way to ensure refrigerant separation with dual-refrigerant circuits is a modification based on Figure 3-23, where only part of the recycling unit circuit is common between the refrigerants.[13] This unit is easier to clear when switching refrigerants, because R-12 can remain in the R-12 condenser.

A sliding door[13] can be used to expose either R-12 or R-134a hand valves and connector fittings, Figures 10-6 and 10-7. This door also permits connections to only one storage container at a time.

The SAE Defrost and Interior Climate Control Committee is writing a standard for refrigerant recovery-recycling equipment for use with both R-12 and R-134a, using common refrigerant circuits. Equipment meeting this standard is expected to meet U.S. EPA regulations for certified recycling equipment for R-12 (current) and R-134a (after November, 1995).

## Recycling Equipment

Recycling equipment must remove moisture and air, separate and measure oil, and have a filter that removes particles and acids in order to meet certification requirements.

Before selecting recycling equipment, the automotive repair shop owner or service technician may wish to assess the particular operation, as well as the following list of equipment options:

- **Filter-drying** — Either single- or multiple-pass filters may be used successfully. Multiple-pass filtering provides the temperature control for better air purging. Locating the drier before the compressor helps keep the compressor oil sump clean.
- **Purging air** — Automatic air-purge methods that circulate the refrigerant for temperature control are preferred. Newer systems, which trap air in a single-pass operation before the refrigerant reaches the storage container, do not require recirculation.
- **Evacuation** — It is convenient to include a deep vacuum pump for removing air and moisture from the air conditioner after repairs have been made. A single- or two-stage rotary vane vacuum pump with at least 1-cfm (0.47-l/s) capacity and capable of pulling a 1000-micron (0.02-psia) vacuum level is suitable for automotive systems. The vacuum pump oil should be changed weekly.

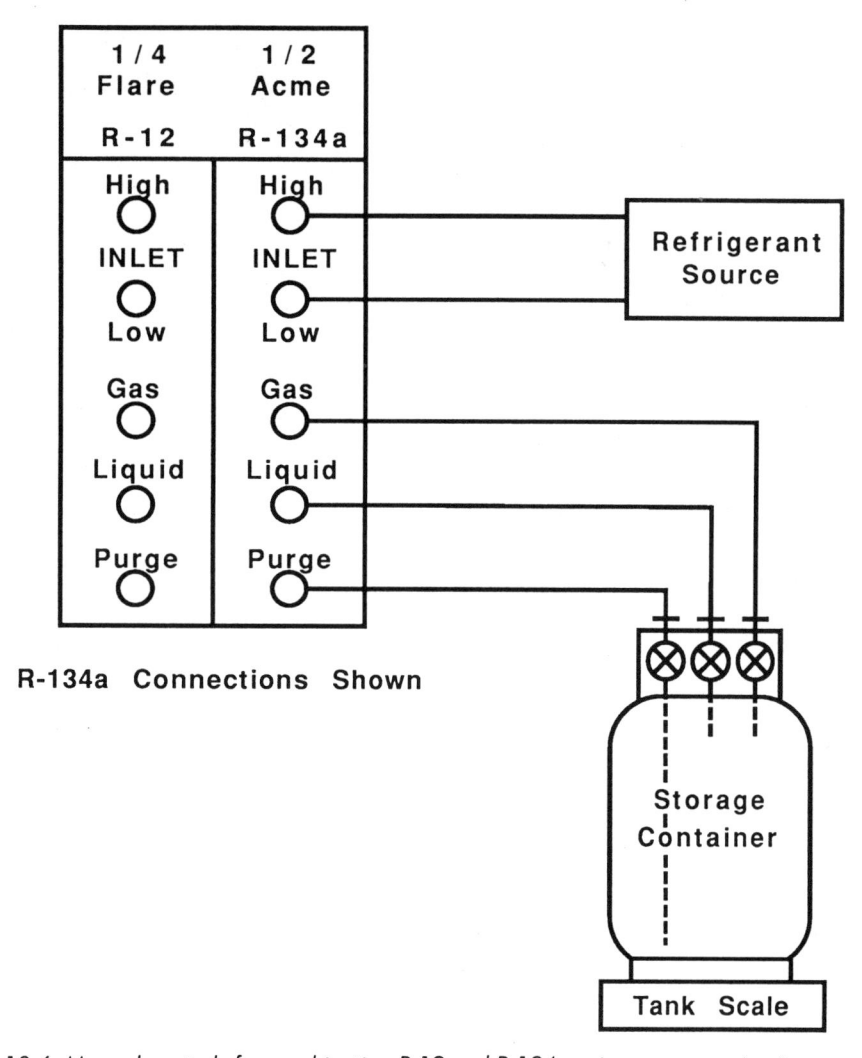

*Figure 10-6. Manual controls for combination R-12 and R-134a using common circuits*

- **Charging** — Many shops prefer charging capability on their refrigerant recycling equipment. A NIST-approved weight scale using a strain gauge load cell can accurately weigh in the charge. The scale also can be used to monitor the refrigerant recovered, but trapped refrigerant in the equipment affects the accuracy of this measurement.
- **Oil charging** — Some recycling equipment models can measure and replenish the amount of oil removed during recovery. The refrigerant may be used to transport the oil into the a/c system.
- **Automatic controls** — Shop owners may wish to select equipment that shuts down upon reaching the final recovery vacuum, on high-side over-pressure, and on tank-filled conditions. Filter-change indicators are also important for continued recycling capabilities.

The Challenge of Recycling Refrigerants

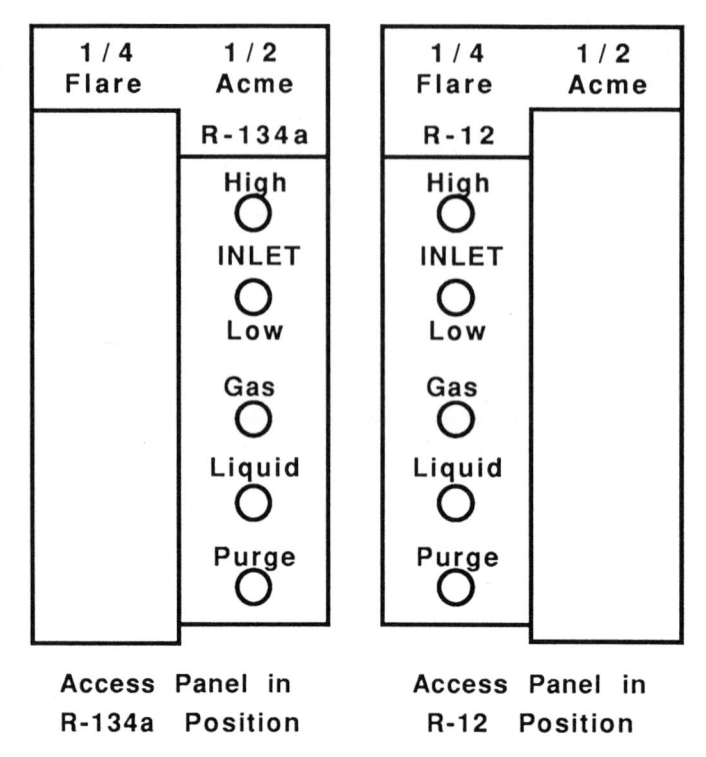

| 1/4 Flare | 1/2 Acme |   | 1/4 Flare | 1/2 Acme |
|---|---|---|---|---|
|  | R-134a |  | R-12 |  |
|  | High ○ |  | High ○ |  |
|  | INLET ○ |  | INLET ○ |  |
|  | Low |  | Low |  |
|  | Gas ○ |  | Gas ○ |  |
|  | Liquid ○ |  | Liquid ○ |  |
|  | Purge ○ |  | Purge ○ |  |
| Access Panel in R-134a Position |  |  | Access Panel in R-12 Position |  |

*Figure 10-7. Manual controls showing sliding door*

## Flushing

During severe service or retrofit, the service technician may choose to thoroughly clean the system and remove the used oil. A flushing adapter kit, Figure 10-8, may be used with the existing recycling unit for this purpose.

The flushing inlet connects to the storage container liquid valve. The flushing outlet connects to the recycling unit inlet through a liquid expansion device, filter, and sight glass. While flushing, the tank liquid valve is opened; the recovery sequence is started; and liquid refrigerant flows through the system or component to be flushed. The flushing filter, regular filter, and oil separator clean the refrigerant before returning it to the storage container.

When the sight glass clears, indicating that the flushing process is finished, the tank liquid valve is closed to allow the recycling unit to remove the refrigerant from the system. Gentle heating or starting the vehicle with a/c off may be required to ensure liquid refrigerant is fully removed.

Evacuate the a/c system for at least 30 minutes after completing the flushing procedure, using the type of vacuum pump described earlier, before charging with new refrigerant and lubricant.

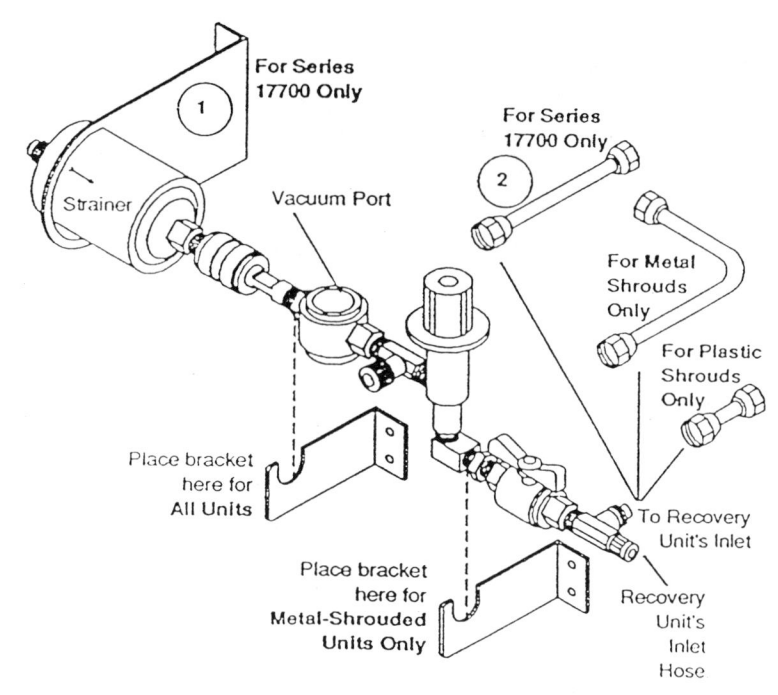

*Figure 10-8. Flushing kit (Courtesy, Robinair Division, SPX Corporation)*

# Refrigerant Identification

Testing for refrigerant types in MVAC systems is necessary because of the following reasons:

- Introduction of R-22 in 14-oz cans on automotive parts shelves at about the same time R-12 cans were prohibited by EPA regulations
- Limited availability of interim refrigerant blends to top off or charge systems that are low on refrigerant (some of these blends are flammable)
- Early phaseout of CFCs (1996), which may require retrofitting a substantial portion of the vehicle fleet
- Migration of used or recycled refrigerant from stationary equipment sectors, where many more refrigerants are used

Charging or topping off a system with some "strange" refrigerant would normally contaminate the refrigerant by more than 10%. A typical vehicle contains 2 to 4 lb (0.9 to 1.8 kg) of refrigerant, which is usually recovered into a tank holding 25 to 45 lb (11.3 to 20.4 kg). If one out of 10 cars is contaminated by 10% of "strange" refrigerant, the tank would still be 99% pure. If the refrigerant is tested before it is recovered into the larger storage container, a 90% purity test to identify R-12, R-134a, or "mixed refrigerant" is considered adequate to protect the used refrigerant supply, Figure 10-9.

The Challenge of Recycling Refrigerants

*Figure 10-9. Automotive refrigerant identifier (Courtesy, Robinair Division, SPX Corporation)*

Due to concerns with R-12 and PAG oil compatibility (see Chapter 7), the tolerance for residual R-12 may be less than 10%. While vehicle manufacturers may recommend compatible lubricants for retrofitting R-12 vehicles to R-134a, it may be appropriate to analyze the refrigerant after retrofitting to ensure that it is relatively pure R-134a. For this application, the refrigerant identifier may need to be capable of measuring a purity level higher than 90%.

## Handling Mixed Refrigerants

Recovery-recycling equipment approved for R-12 or R-134a MVAC use should not be used to recover refrigerant that has been tested and identified as mixed refrigerant for the following reasons:

- **Cross-contamination of the used-refrigerant supply** — If R-12 units are to be used only for R-12 and R-134a units only for R-134a, neither should be used for contaminated or mixed refrigerant.
- **Flammability** — Some interim replacement refrigerant blends are themselves flammable, or may become flammable if substantial leakage occurs. The hermetic compressor in the recycling equipment or other electrical components could ignite flammable refrigerant.

Tests performed demonstrated a car bursting into flames with a reasonable charge of refrigerant and a reasonable ignition source. No injuries have been reported that were caused by actual use of flammable refrigerants in MVACs.

The subject of flammable refrigerants is controversial in the U.S.; flammable refrigerants have not been used in MVACs since their introduction in the 1940s. Experts in industry, government, and testing laboratories are working on standardized tests, ignition sources, and risk-assessment techniques. Recovery-recycling equipment manufacturers have not examined their equipment to the extent required to confidently recommend its use for flammable refrigerants.

Equipment used for the purpose of recovering mixed refrigerants should be constructed using air power and manual valves, explosion-proof electrical motor and components, or similar construction. A schematic of an air-motor-powered "scavenger unit"[13] is shown in Figure 10-10.

## Federal Equipment Requirements

The entire text of U.S. EPA Section 609 Regulations (Part 82 Subpart B) is included in Figure 10-11. Approved refrigerant recycling equipment for R-12 had to be certified to EPA requirements by January 1, 1993, Figure 10-12. Effective November 15, 1995, approved recycling equipment must also be used for R-134a or any other replacement refrigerant.

## The Challenge of Recycling Refrigerants

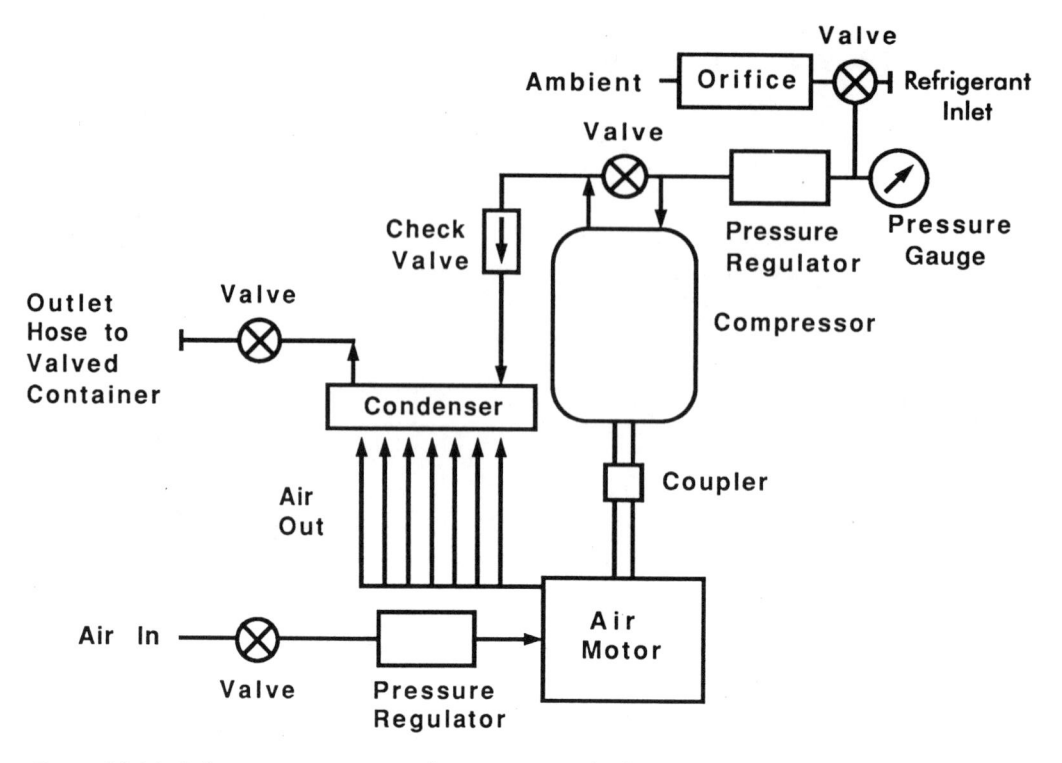

*Figure 10-10. Refrigerant recovery unit for contaminated refrigerants*

List of Subjects in 40 CFR Part 82

Administrative practice and procedure, Air pollution control, Reporting and recordkeeping requirements, Stratospheric ozone layer.

Dated: April 23, 1993.

**Carol M. Browner,**

*Administrator.*

For the reasons set out in the preamble, 40 CFR part 82 is amended as follows:

**PART 82—PROTECTION OF STRATOSPHERIC OZONE**

1. Authority: The authority citation for part 82 continues to read as follows:

**Authority:** 42 U.S.C. 7414, 7601, 7671–7671q.

2. Part 82 is amended by adding subpart F to read as follows:

**Subpart F—Recycling and Emissions Reduction**

Sec.
- 82.150 Purpose and scope.
- 82.152 Definitions.
- 82.154 Prohibitions.
- 82.156 Required practices.
- 82.158 Standards for recycling and recovery equipment.
- 82.160 Approved equipment testing organizations.
- 82.161 Technician certification.
- 82.162 Certification by owners of recovery and recycling equipment.
- 82.164 Reclaimer certification.
- 82.166 Reporting and recordkeeping requirements.

*Figure 10-11. Federal EPA Section 609 regulations (Source: EPA)*

2. Part 82 is amended by designating the existing sections and appendices as subpart A and by adding a new subpart B to read as follows:

**Subpart A—Production and Consumption Controls**

\* \* \* \* \*

**Subpart B—Servicing of Motor Vehicle Air Conditioners**

Sec.
82.30   Purpose and Scope.
82.32   Definitions.
82.34   Prohibitions.
82.36   Approved Refrigerant Recycling Equipment.
82.38   Approved Independent Standards Testing Organizations.
82.40   Technician Training and Certification.
82.42   Certification, Recordkeeping and Public Notification Requirements.

Appendix A to Part 82 Subpart B—Standard for Recycle/Recover Equipment

Appendix B to Part 82 Subpart B—Standard for Recover Equipment [Reserved]

### § 82.30  Purpose and scope.

(a) The purpose of these regulations is to implement section 609 of the Clean Air Act, as amended (Act) regarding the servicing of motor vehicle air conditioners.

(b) These regulations apply to any person performing service on a motor vehicle for consideration when this service involves the refrigerant in the motor vehicle air conditioner.

### § 82.32  Definitions.

(a) *Approved Independent Standards Testing Organization* means any organization which has applied for and received approval from the Administrator pursuant to § 82.38.

(b) *Approved Refrigerant Recycling Equipment* means equipment certified by the Administrator or an organization approved under § 82.38 as meeting either one of the standards in § 82.36. Such equipment extracts and recycles refrigerant or extracts refrigerant for recycling on-site or reclamation off-site.

(c) *Motor vehicle* as used in this subpart means any vehicle which is self-propelled and designed for transporting persons or property on a street or highway, including but not limited to passenger cars, light duty vehicles, and heavy duty vehicles. This definition does not include a vehicle where final assembly of the vehicle has not been completed by the original equipment manufacturer.

(d) *Motor vehicle air conditioners* means mechanical vapor compression refrigeration equipment used to cool the driver's or passenger's compartment of any motor vehicle. This definition is not intended to encompass the hermetically sealed refrigeration systems used on motor vehicles for refrigerated cargo and the air conditioning systems on passenger buses using HCFC-22 refrigerant.

(e) *Properly using* means using equipment in conformity with Recommended Service Procedure for the Containment of R-12 (CFC-12) set forth in appendix A to this subpart. In addition, this term includes operating the equipment in accordance with the manufacturer's guide to operation and maintenance and using the equipment only for the controlled substance for which the machine is designed. For equipment that extracts and recycles refrigerant, properly using also means to recycle refrigerant before it is returned to a motor vehicle air conditioner. For equipment that only recovers refrigerant, properly using includes the requirement to recycle the refrigerant on-site or send the refrigerant off-site for reclamation. Refrigerant from reclamation facilities that is used for the purpose of recharging motor vehicle air conditioners must be at or above the standard of purity developed by the Air-conditioning and Refrigeration Institute (ARI 700–88) (available at 4301 North Fairfax Drive, Suite 425, Arlington, Virginia 22203) in effect as of November 15, 1990. Refrigerant may be recycled off-site only if the refrigerant is extracted using recover only equipment, and is subsequently recycled off-site by equipment owned by the person that owns both the recover only equipment and owns or operates the establishment at which the refrigerant was extracted. In any event, approved equipment must be used to extract refrigerant prior to performing any service during which discharge of refrigerant from the motor vehicle air conditioner can reasonably be expected. Intentionally venting or disposing of refrigerant to the atmosphere is an improper use of equipment.

(f) *Refrigerant* means any class I or class II substance used in a motor vehicle air conditioner. Class I and class II substances are listed in part 82, subpart A, appendix A. Effective November 15, 1995, refrigerant shall also include any substitute substance.

(g) *Service for consideration* means being paid to perform service, whether it is in cash, credit, goods, or services. This includes all service except that done for free.

(h) *Service involving refrigerant* means any service during which discharge or release of refrigerant from the motor vehicle air conditioner to the atmosphere can reasonably be expected to occur.

### § 82.34  Prohibitions.

(a) Effective August 13, 1992, no person repairing or servicing motor vehicles for consideration may perform any service on a motor vehicle air conditioner involving the refrigerant for such air conditioner

(1) Without properly using equipment approved pursuant to § 82.36; and

(2) Unless such person has been properly trained and certified by a technician certification program approved by the Administrator pursuant to § 82.40.

The requirements of this paragraph do not apply until January 1, 1993 for small entities who certify to the Administrator in accordance with § 82.42(a)(2).

(b) Effective November 15, 1992, no person may sell or distribute, or offer for sale or distribution, any class I or class II substance that is suitable for use as a refrigerant in motor vehicle air-conditioner and that is in a container which contains less than 20 pounds of such refrigerant to any person unless that person is properly trained and certified under § 82.40 or intended the containers for resale only, and so certifies to the seller under § 82.42(b)(4).

(c) No technician training programs may issue certificates unless the program complies with all of the standards in § 82.40(a).

### § 82.36  Approved refrigerant recycling equipment.

(a)(1) Refrigerant recycling equipment must be certified by the Administrator or an independent standards testing organization approved by the Administrator under § 82.38 to meet the following standard:

(2) Equipment that recovers and recycles refrigerant must meet the standards set forth in appendix A to this subpart (Recommended Service Procedure for the Containment of R-12, Extraction and Recycle Equipment for Mobile Automotive Air-Conditioning Systems, and Standard of Purity for Use in Mobile Air Conditioning Systems).

(b) Refrigerant recycling equipment purchased before September 4, 1991 that has not been certified under paragraph (a) of this section shall be considered approved if the equipment is substantially identical to equipment certified under paragraph (a) of this section. Equipment manufacturers or owners may request a determination by the Administrator by submitting an application and supporting documents which indicate that the equipment is substantially identical to approved equipment to: MVACs Recycling Program Manager, Stratospheric Ozone Protection Branch (6202-J), U.S.

Environmental Protection Agency, 401 M Street, SW., Washington, DC 20460, Attn. Substantially Identical Equipment Review.

Supporting documents must include process flow sheets, lists of components and any other information which would indicate that the equipment is capable of cleaning the refrigerant to the standards in appendix A. Authorized representatives of the Administrator may inspect equipment for which approval is being sought and request samples of refrigerant that has been extracted and/or recycled using the equipment. Equipment which fails to meet appropriate standards will not be considered approved.

(c) The Administrator will maintain a list of approved equipment by manufacturer and model. Persons interested in obtaining a copy of the list should send written inquiries to the address in paragraph (b) of this section.

*Figure 10-11. Federal EPA Section 609 regulations — continued*

## § 82.38 Approved independent standards testing organizations.

(a) Any independent standards testing organization may apply for approval by the Administrator to certify equipment as meeting the standards in appendix A to this subpart. The application shall be sent to: MVACs Recycling Program Manager, Stratospheric Ozone Protection Branch (6202–J), U.S. Environmental Protection Agency, 401 M Street, SW., Washington, DC 20460.

(b) Applications for approval must document the following:

(1) That the organization has the capacity to accurately test whether refrigerant recycling equipment complies with the applicable standards. In particular, applications must document:

(i) The equipment present at the organization that will be used for equipment testing;

(ii) The expertise in equipment testing and the technical experience of the organization's personnel;

(iii) Thorough knowledge of the standards as they appear in appendix A of this subpart; and

(iv) The test procedures to be used to test equipment for compliance with applicable standards, and why such test procedures are appropriate for that purpose.

(2) That the organization has no conflict of interest and will receive no financial benefit based on the outcome of certification testing; and

(3) That the organization agrees to allow the Administrator access to verify the information contained in the application.

(c) If approval is denied under this section, the Administrator shall give written notice to the organization setting forth the basis for his or her determination.

(d) If at any time an approved independent standards testing organization is found to be conducting certification tests for the purposes of this subpart in a manner not consistent with the representations made in its application for approval under this section, the Administrator reserves the right to revoke approval.

## § 82.40 Technician training and certification.

(a) Any technician training and certification program may apply for approval, in accordance with the provisions of this paragraph, by submitting to the Administrator at the address in § 82.38 (a) verification that the program meets all of the following standards:

(1) *Training.* Each program must provide adequate training, through one or more of the following means: on-the-job training, training through self-study of instructional material, or on-site training involving instructors, videos or a hands-on demonstration.

(2) *Test Subject Material.* The certification tests must adequately and sufficiently cover the following:

(i) The standards established for the service and repair of motor vehicle air conditioners as set forth in Appendix A to this subpart. These standards relate to the recommended service procedures for the containment of refrigerant, extraction and recycle equipment, and the standard of purity for refrigerant in motor vehicle air conditioners.

(ii) Anticipated future technological developments, such as the introduction of HFC–134a in new motor vehicle air conditioners.

(iii) The environmental consequences of refrigerant release and the adverse effects of stratospheric ozone layer depletion.

(iv) As of August 13, 1992, the requirements imposed by the Administrator under § 609 of the Act.

(3) *Test Administration.* Completed tests must be graded by an entity or individual who receives no benefit based on the outcome of testing; a fee may be charged for grading. Sufficient measures must be taken at the test site to ensure that tests are completed honestly by each technician. Each test must provide a means of verifying the identification of the individual taking the test. Programs are encouraged to make provisions for non-English speaking technicians by providing tests in other languages or allowing the use of a translator when taking the test. If a translator is used, the certificate

# The Challenge of Recycling Refrigerants

received must indicate that translator assistance was required.

(4) *Proof of Certification.* Each certification program must offer individual proof of certification, such as a certificate, wallet-sized card, or display card, upon successful completion of the test. Each certification program must provide a unique number for each certified technician.

(b) In deciding whether to approve an application, the Administrator will consider the extent to which the applicant has documented that its program meets the standards set forth in this section. The Administrator reserves the right to consider other factors deemed relevant to ensure the effectiveness of certification programs. The Administrator may approve a program which meets all of the standards in paragraph (a) of this section except test administration if the program, when viewed as a whole, is at least as effective as a program that does meet all the standards. Such approval shall be limited to training and certification conducted before August 13, 1992. If approval is denied under this section, the Administrator shall give written notice to the program setting forth the basis for his determination.

(c) *Technical Revisions.* Directors of approved certification programs must conduct periodic reviews of test subject material and update the material based upon the latest technological developments in motor vehicle air conditioner service and repair. A written summary of the review and any changes made must be submitted to the Administrator every two years.

(d) *Recertification.* The Administrator reserves the right to specify the need for technician recertification at some future date, if necessary.

(e) If at any time an approved program is conducted in a manner not consistent with the representations made in the application for approval of the program under this section, the Administrator reserves the right to revoke approval.

(f) Authorized representatives of the Administrator may require technicians to demonstrate on the business entity's premises their ability to perform proper procedures for recovering and/or recycling refrigerant. Failure to demonstrate or failure to properly use the equipment may result in revocation of the technician's certificate by the Administrator. Technicians whose certification is revoked must be recertified before servicing or repairing any motor vehicle air conditioners.

**§ 82.42 Certification, recordkeeping and public notification requirements.**

(a) *Certification requirements.* (1) No later than January 1, 1993, any person repairing or servicing motor vehicle air conditioners for consideration shall certify to the Administrator that such person has acquired, and is properly using, approved equipment and that each individual authorized to use the equipment is properly trained and certified. Certification shall take the form of a statement signed by the owner of the equipment or another responsible officer and setting forth:

(i) The name of the purchaser of the equipment;

(ii) The address of the establishment where the equipment will be located; and

(iii) The manufacturer name and equipment model number, the date of manufacture, and the serial number of the equipment. The certification must also include a statement that the equipment will be properly used in servicing motor vehicle air conditioners, that each individual authorized by the purchaser to perform service is properly trained and certified in accordance with § 82.40, and that the information given is true and correct. The certification should be sent to: MVACs Recycling Program Manager, Stratospheric Ozone Protection Branch (6202–J), U.S. Environmental Protection Agency, 401 M Street, SW., Washington, DC 20460.

(2) The prohibitions in § 82.34(a) shall be effective as of January 1, 1993 for persons repairing or servicing motor vehicle air conditioners for consideration at an entity which performed service on fewer than 100 motor vehicle air conditioners in calendar year 1990, but only if such person so certifies to the Administrator no later than August 13, 1992. Persons must retain adequate records to demonstrate that the number of vehicles serviced was fewer than 100.

*Figure 10-11. Federal EPA Section 609 regulations — continued*

(3) Certificates of compliance are not transferable. In the event of a change of ownership of an entity which services motor vehicle air conditioners for consideration, the new owner of the entity shall certify within thirty days of the change of ownership pursuant to § 82.42(a)(1).

(b) *Recordkeeping requirements.* (1) Any person who owns approved refrigerant recycling equipment certified under § 82.36(a)(2) must maintain records of the name and address of any facility to which refrigerant is sent.

(2) Any person who owns approved refrigerant recycling equipment must retain records demonstrating that all persons authorized to operate the equipment are currently certified under § 82.40.

(3) Any person who sells or distributes any class I or class II substance that is suitable for use as a refrigerant in a motor vehicle air conditioner and that is in a container of less than 20 pounds of such refrigerant must verify that the purchaser is properly trained and certified under § 82.40. The seller must have a reasonable basis for believing that the information presented by the purchaser is accurate. The only exception to these requirements is if the purchaser is purchasing the small containers for resale only. In this case, the seller must obtain a written statement from the purchaser that the containers are for resale only and indicate the purchasers name and business address. Records required under this paragraph must be retained for a period of three years.

(4) All records required to be maintained pursuant to this section must be kept for a minimum of three years unless otherwise indicated. Entities which service motor vehicle air conditioners for consideration must keep these records on-site.

(5) All entities which service motor vehicle air conditioners for consideration must allow an authorized representative of the Administrator entry onto their premises (upon presentation of his or her credentials) and give the authorized representative access to all records required to be maintained pursuant to this section.

(c) *Public Notification.* Any person who conducts any retail sales of a class I or class II substance that is suitable for use as a refrigerant in a motor vehicle air conditioner, and that is in a container of less than 20 pounds of refrigerant, must prominently display a sign where sales of such containers occur which states:

"It is a violation of federal law to sell containers of Class I and Class II refrigerant of less than 20 pounds of such refrigerant to anyone who is not properly trained and certified to operate approved refrigerant recycling equipment."

## Appendix A to Subpart B—Standard for Recycle/Recover Equipment

### Standard of Purity for Use in Mobile Air-Conditioning Systems

*Foreword*

Due to the CFC's damaging effect on the ozone layer, recycle of CFC-12 (R-12) used in mobile air-conditioning systems is required to reduce system venting during normal service operations. Establishing recycle specifications for R-12 will assure that system operation with recycled R-12 will provide the same level of performance as new refrigerant.

Extensive field testing with the EPA and the auto industry indicate that reuse of R-12 removed from mobile air-conditioning systems can be considered, if the refrigerant is cleaned to a specific standard. The purpose of this standard is to establish the specific minimum levels of R-12 purity required for recycled R-12 removed from mobile automotive air-conditioning systems.

*1. Scope*

This information applies to refrigerant used to service automobiles, light trucks, and other vehicles with similar CFC-12 systems. Systems used on mobile vehicles for refrigerated cargo that have hermetically sealed, rigid pipe are not covered in this document.

*2. References*

SAE J1989, Recommended Service Procedure for the Containment of R-12
SAE J1990, Extraction and Recycle Equipment for Mobile Automotive Air-Conditioning Systems
ARI Standard 700-88

## 3. Purity Specification

The refrigerant in this document shall have been directly removed from, and intended to be returned to, a mobile air-conditioning system. The contaminants in this recycled refrigerant 12 shall be limited to moisture, refrigerant oil, and noncondensable gases, which shall not exceed the following level:

3.1 *Moisture:* 15 ppm by weight.
3.2 *Refrigerant Oil:* 4000 ppm by weight.
3.3 *Noncondensable Gases (air):* 330 ppm by wright.

## 4. Refrigeration Recycle Equipment Used in Direct Mobile Air-Conditioning Service Operations Requirement

4.1 The equipment shall meet SAE J1990, which covers additional moisture, acid, and filter requirements.
4.2 The equipment shall have a label indicating that it is certified to meet this document.

## 5. Purity Specification of Recycled R-12 Refrigerant Supplied in Containers From Other Recycle Sources

Purity specification of recycled R-12 refrigerant supplied in containers from other recycle sources, for service of mobile air-conditioning systems, shall meet ARI Standard 700-88 (Air Conditioning and Refrigeration Institute).

## 6. Operation of the Recycle Equipment

This shall be done in accordance with SAE J1989.

*Rationale*

Not applicable.

*Relationship of SAE Standard to ISO Standard*

Not applicable.

*Reference Section*

SAE. J1989. Recommended Service Procedure for the Containment of R-12

Figure 10-11. Federal EPA Section 609 regulations — concluded

## MVAC RECOVER/RECYCLE OR RECOVER EQUIPMENT CERTIFICATION FORM

**1** Name of Establishment

Street

City, State, Zip

(Area Code) Telephone Number

**2** Name of Equipment Manufacturer and Model Number

Serial Number(s) / Year

**3** I certify that I have acquired approved recover/recycle equipment under Section 609 of the Clean Air Act. I certify that only properly trained and certified technicians operate the equipment and that the information given above is true and correct.

Signature of Owner/Operator           Date

Name (Please Print)           Title

### *MOBILE AIR CONDITIONING RECOVER/RECYCLE EQUIPMENT CERTIFICATION FORM INSTRUCTIONS*

Motor vehicle recover/recycle equipment must be acquired by January 1, 1992 and certified to EPA on or before January 1, 1993 under Section 609 of the Clean Air Act. To certify your equipment, please complete the above form according to the following instructions and *mail to EPA at the following address:*

**MVAC Recycling Program Manager, Stratospheric Ozone Protection Branch, (6202-J), 401 M Street, S.W. Washington, D.C. 20460.**

**1** Please provide the name, address and telephone number of the establishment where the recover/recycle equipment is located.

**2** Please provide the name brand, model number, year, and serial number(s) of the recover/recycle equipment acquired for us at the above establishment.

**3** The certification statement must be signed by the person who has acquired the recover/recycle equipment (the person may be the owner of the establishment or another responsible officer). The person who signs is certifying that they have acquired the equipment, that each individual authorized to use the equipment is properly trained and certified, and that the information provided is true and correct.

*Figure 10-12. Automotive recycling equipment certification form (Source: EPA)*

## Summary

The SAE Interior Climate Control Committee and the motor vehicle industry have set standards for recovery-recycling equipment, refrigerant purity, and service procedures for separate R-12 and R-134a equipment. They have also developed recommended service practices for retrofitting R-12 vehicles to R-134a.

Within the SAE requirements, design choices are available. Some equipment offers additional capabilities, such as evacuation, refrigerant charging, and oil charging. Flushing kits are available for use with existing recycling units and for completely removing oil and cleaning severely contaminated systems. Refrigerant identifiers are available to identify R-12 and R-134a that is pure enough for reuse.

Because of cross-contamination and flammability issues, existing R-12 or R-134a recycling units should not be used to recover refrigerants tested and found to be contaminated. Instead, air-powered or explosion-proof equipment (or equivalent) should be used for these "scavenger" purposes.

Federal requirements for owning and using approved (certified) recycling equipment are based on the SAE standards.

## References

[1] SAE J1990, 1989, *Extraction and Recycle Equipment for Mobile Automotive Air-Conditioning Systems.*

[2] SAE J1991, 1989, *Standard of Purity for Use in Mobile Air-Conditioning Systems.*

[3] SAE J1989, 1989, *Recommended Service Procedure for the Containment of R-12.*

[4] SAE J2210, 1991, *HFC-134a Recycling Equipment for Mobile Air-Conditioning Systems.*

[5] SAE J2099, 1991, *Standard of Purity for Recycled HFC-134a for Use in Mobile Air-Conditioning Systems.*

[6] SAE J2211, 1991, *Recommended Service Procedure for the Containment of HFC-134a.*

[7] SAE J639, 1994, *Safety and Containment of Refrigerant for Mechanical Vapor Compression Systems Used for Mobile Air-Conditioning Systems.*

[8] SAE J2197, 1992, *HFC-134a (R-134a) Service Hose Fittings for Automotive Air-Conditioning Service Equipment.*

[9] SAE J2196, 1992, *Service Hose for Automotive Air Conditioning.*

[10] SAE J1660, 1993, *Fittings and Labels for Retrofit of CFC-12 (R-12) Mobile Air-Conditioning Systems to HFC-134a (R-134a).*

[11] SAE J1661, 1993, *Procedure for Retrofitting CFC-12 (R-12) Mobile Air-Conditioning Systems to HFC-134a (R-134a)*.

[12] Kenneth W. Manz, "Survey of Residual Refrigerant 12 in Vehicle Air Conditioners Retrofit to R-134a." International CFC and Halon Alternatives Conference in Washington D.C., Sept., 1992.

[13] U.S. Patent Pending, Robinair Division, SPX Corporation.

Chapter Eleven

# Service Procedures

This chapter will cover the general steps of the service procedure and highlight changes brought about by stratospheric ozone depletion, the resulting phaseout of CFC and HCFC refrigerants, and recycling practices to minimize the amount of refrigerant released to the atmosphere.

## Connecting

For stationary applications, the system service ports remain unchanged while in motor vehicles; special non-threaded service ports are provided for R-134a systems. Service hoses should include shutoff valves to prevent the connect-and-purge method used in the past, Figure 11-1.

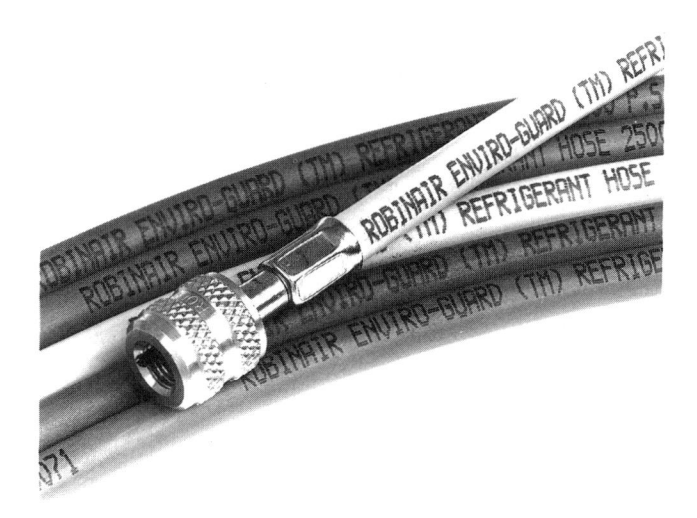

*Figure 11-1. Low-permeation hose with automatic shutoff valve (Courtesy, Robinair Division, SPX Corporation)*

# The Challenge of Recycling Refrigerants

In general, hoses are color-coded red for high-side connection, blue for low-side connection, and yellow for general purpose.

Low-permeation hoses (nylon-barrier construction) are required as part of SAE and ARI recovery-recycling equipment certification programs. For stationary applications and R-12 motor vehicle systems, purchase hoses marked "SAE J2196 R-12." For motor vehicle R-134a systems, purchase hoses marked "SAE J2196 R-134a." Removing valve cores with a tool like the one in Figure 11-2 will increase the refrigerant recovery and evacuation rates.

*Figure 11-2. Valve core-removing tube (Courtesy, Robinair Division, SPX Corporation)*

## Diagnosing

The manifold gauge, Figure 11-3, is the primary diagnostic tool. The manifold provides a means of connecting to both the system service ports and to service equipment. It further provides selective access to high- and low-side ports through the shutoff valves which, like the hoses, are color-coded red for high side and blue for low side.

The manifold pressure gauges indicate system operating pressures and evaporating and condensing temperatures. The gauges are useful for observing system cycling times.

System temperature measurements supplement the manifold gauge set in determining proper system operation. A pocket thermometer, Figure 11-4, may be used to measure evaporator discharge air temperature, while a thermocouple tester, Figure 11-5, may be used to measure various line temperatures in the refrigeration system.

Electrical instruments, such as the multimeter shown in Figure 11-6 or the compressor analyzer shown in Figure 11-7, are used to diagnose electrical and control malfunctions.

Service Procedures

*Figure 11-3. Manifold gauge set (Courtesy, Robinair Division, SPX Corporation)*

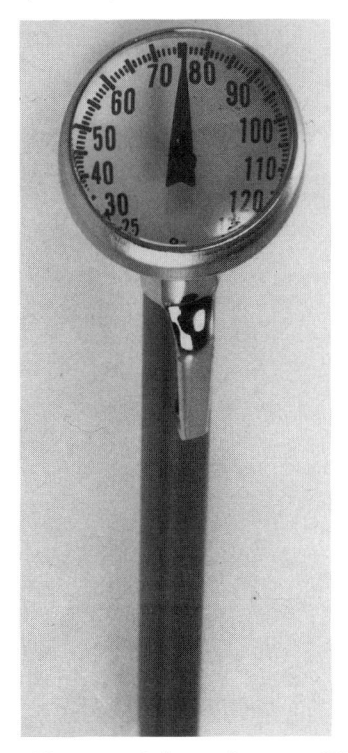

*Figure 11-4. Pocket thermometer (Courtesy, Robinair Division, SPX Corporation)*

The Challenge of Recycling Refrigerants

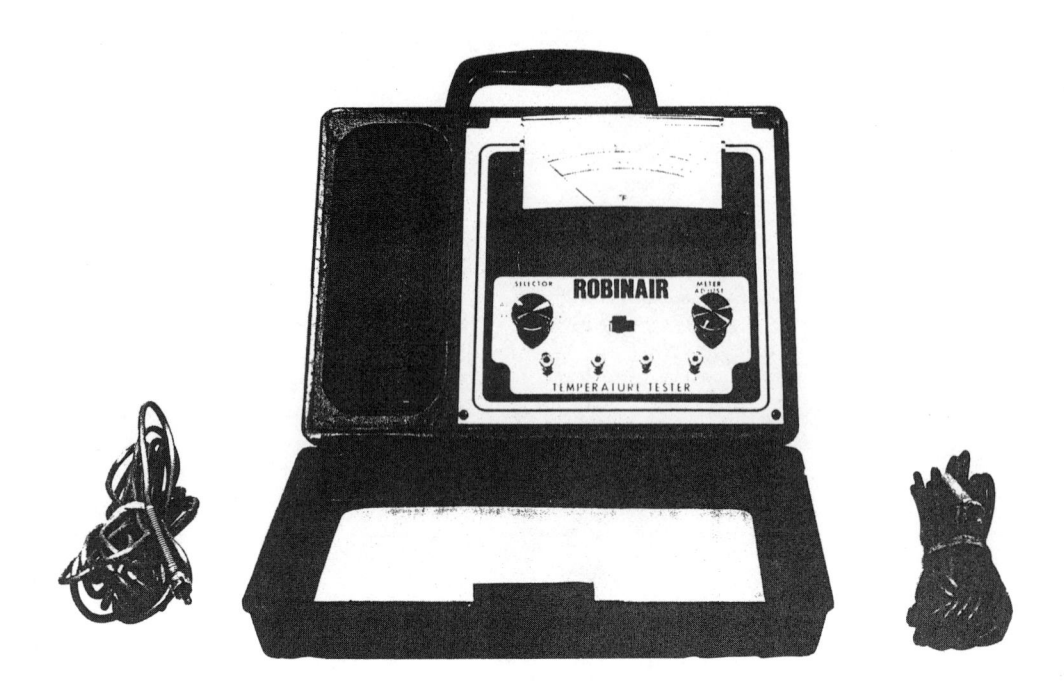

Figure 11-5. Thermocouple temperature tester (Courtesy, Robinair Division, SPX Corporation)

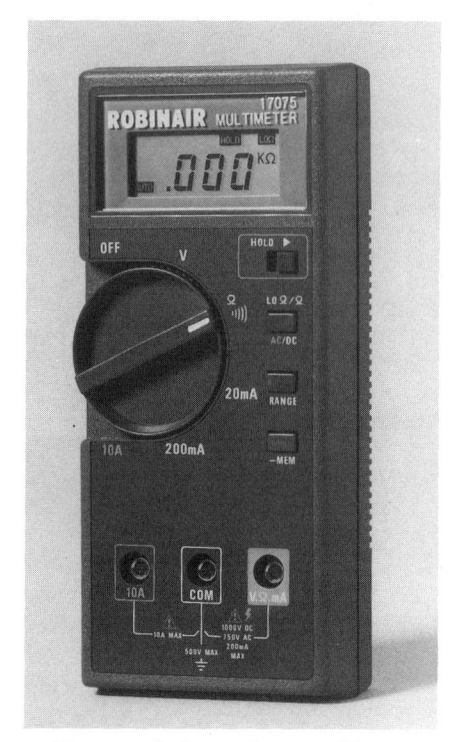

Figure 11-6. Electrical multimeter (Courtesy, Robinair Division, SPX Corporation)

Service Procedures

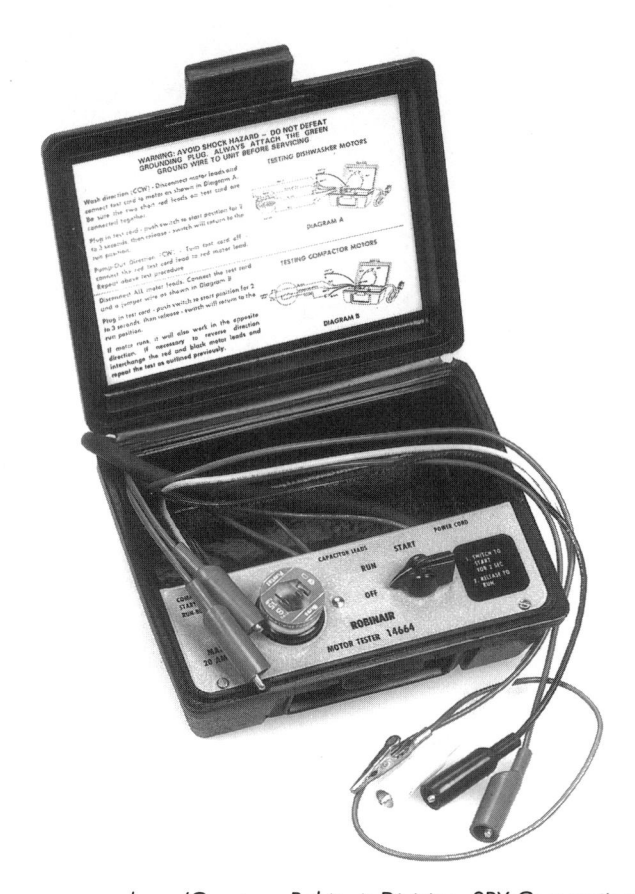

*Figure 11-7. Compressor analyzer (Courtesy, Robinair Division, SPX Corporation)*

Refrigerant conservation practices (finding and fixing leaks) will be increasingly emphasized in regulations and manufacturer recommendations. An electronic leak detector like the one shown in Figure 11-8 may be used for this purpose.

Since mixed refrigerants are the most significant potential problem brought on by recycling and the introduction of interim and replacement refrigerants, screening refrigerants before recovery will be required whenever the refrigerant composition is suspect. If refrigerant from every system is screened, the capability of the refrigerant identifier need not be as accurate as one used to analyze large batches after consolidation, especially if the refrigerant is to be sold. A refrigerant identifier suitable for screening R-12 and R-134a for motor vehicle system usage was shown in Figure 10-9.

Finally, system monitors are increasingly used in food refrigeration and larger systems to alert owners to system malfunctions. The amount of instrumentation and level of sophistication varies.

The Challenge of Recycling Refrigerants

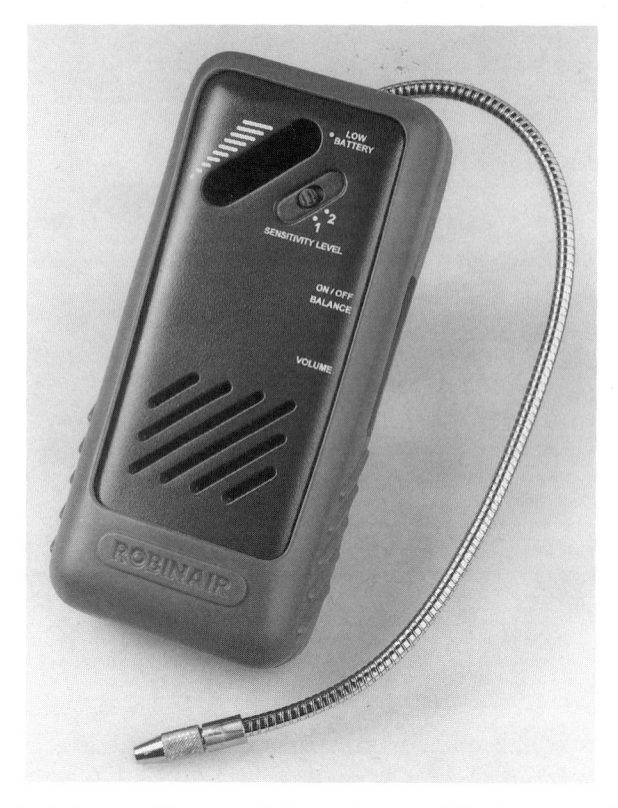

*Figure 11-8. Electronic leak detector (Courtesy, Robinair Division, SPX Corporation)*

## Removing Refrigerant

Because systems are pressurized, refrigerant must be isolated or removed prior to making system repairs. The past practice of releasing refrigerant to the atmosphere is unacceptable due to ozone depletion and global warming concerns and is prohibited in the U.S. and many other countries.

If the system has isolation valves, only a portion of refrigerant in the section to be repaired must be recovered. The recovery-recycling equipment and the storage container used to remove the refrigerant should be appropriate for the refrigerant and for the method of processing the refrigerant before reuse (see Chapters 8 and 9 for stationary systems and Chapter 10 for motor vehicle systems).

Units for recovery with oil separation, for on-site recycling as companion to recovery equipment, for motor vehicle recovery-recycling with recharge, and for central recycling are shown in Figures 11-9 through 11-12, respectively.

Service Procedures

*Figure 11-9. Portable recovery unit with oil separator (Courtesy, Robinair Division, SPX Corporation)*

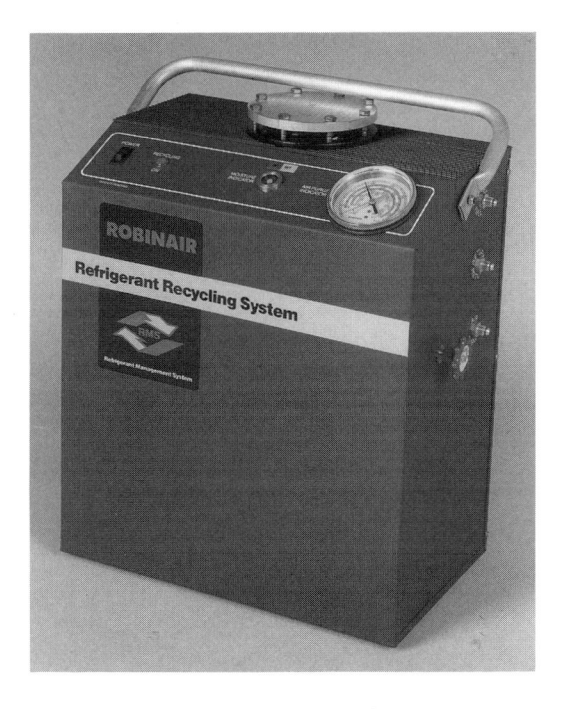

*Figure 11-10. Portable on-site recycling unit (Courtesy, Robinair Division, SPX Corporation)*

167

The Challenge of Recycling Refrigerants

*Figure 11-11. Motor vehicle recovery, recycle, recharge unit (Courtesy, Robinair Division, SPX Corporation)*

*Figure 11-12. Central recycling unit (Courtesy, Robinair Division, SPX Corporation)*

# Service Procedures

## Repairing

The steps of fixing, cleaning, and leak checking are all included under "repairing." All problems revealed when diagnosing should be fixed. If appropriate, inefficient air purgers on centrifugal chillers should be replaced and isolation valves should be installed.

While much attention has been given to the purity of refrigerant used to charge systems after repair, too little attention has been given to impurities left in the system. Most of the lubricant, containing much of the acid, particulate, and chloride, remains in the system after the refrigerant has been recovered. Draining oil, replacing the liquid line filter-drier, adding a suction line filter-drier, and cleaning or replacing the expansion device are common steps to clean up contaminated systems[1].

Acid test kits, Figure 11-13, sight glass moisture indicators, Figure 11-14, and drawn oil samples for laboratory analysis may serve to guide the extent of the clean-up. In some cases, blowing out portions of systems with nitrogen, flushing systems with refrigerant, or wiping down parts or components may be required.

After repairs have been completed, it is necessary to leak check the system to ensure that proper connections and repairs have been accomplished. This can be done by introducing a vapor charge of refrigerant or special leak test mixture such as 5% R-22 and 95% nitrogen. Test pressure should be limited to 10 psig (69 kPa) for low-pressure systems and to manufacturer recommendations for all systems.

## Evacuating

After the system is repaired, any air and moisture that entered while the system was opened must be removed. Rotary vane vacuum pumps, such as the 1.2-cfm (0.57-l/s) single-stage-type shown in Figure 11-15 and the 6-cfm (2.8-l/s) two-stage pump shown in Figure 11-16, are used for this purpose. Pulling from both high- and low-side service ports shortens the time required to reach acceptable vacuum levels. Hoses of larger diameter and shorter length also shorten evacuating time.

The graph in Figure 11-17 shows the general pumpdown curve when evacuating a system. The vacuum pump starts to pump when the system is at atmospheric pressure. Connecting and starting the pump when the system is pressurized could damage the vacuum pump.

The pump removes air throughout the evacuation process but removes water droplets clinging to the walls or moisture from the lubricant only when the gauge at the vacuum pump reads between 500 and 1,000 microns (0.01 and 0.02 psia). Electronic micron gauges, such as the ones shown in Figures 11-18 and 11-19, should be used to ensure that proper vacuum levels have been achieved. When evacuation is completed, it is good practice to valve off the pump and check for pressure rise in the system.

# The Challenge of Recycling Refrigerants

*Figure 11-13. Acid test kit (Courtesy, Sporlan Valve Company)*

*Figure 11-14. Moisture indicator (Courtesy, Sporlan Valve Company)*

Service Procedures

*Figure 11-15. Single-stage vacuum pump (Courtesy, Robinair Division, SPX Corporation)*

*Figure 11-16. Two-stage vacuum pump (Courtesy, Robinair Division, SPX Corporation)*

# The Challenge of Recycling Refrigerants

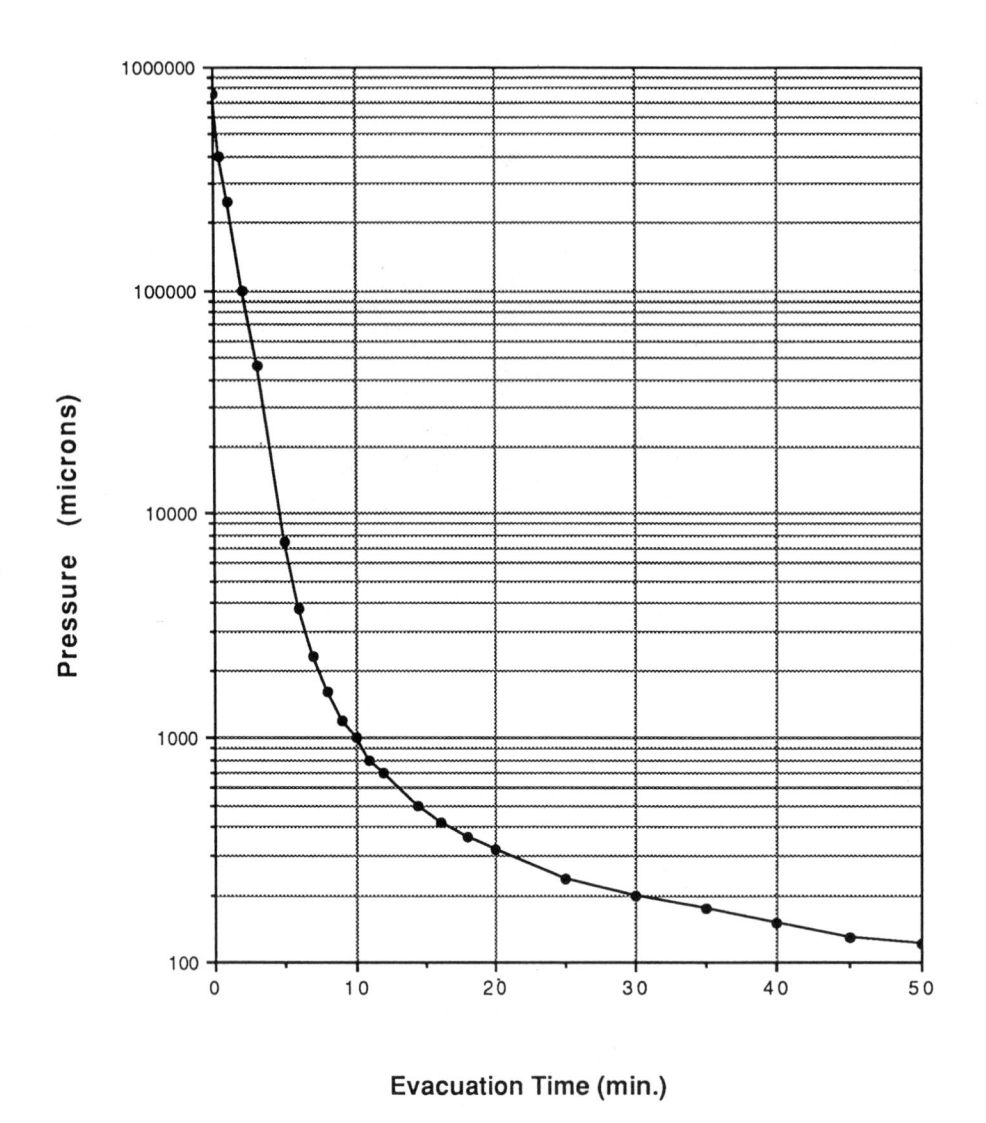

*Figure 11-17. Vacuum pumpdown curve (Courtesy, Robinair Division, SPX Corporation)*

Service Procedures

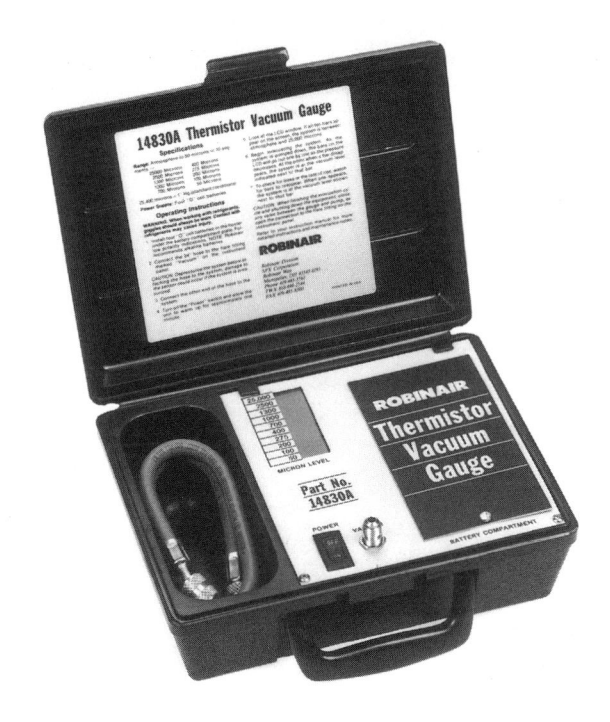

*Figure 11-18. Digital vacuum gauge (Courtesy, Robinair Division, SPX Corporation)*

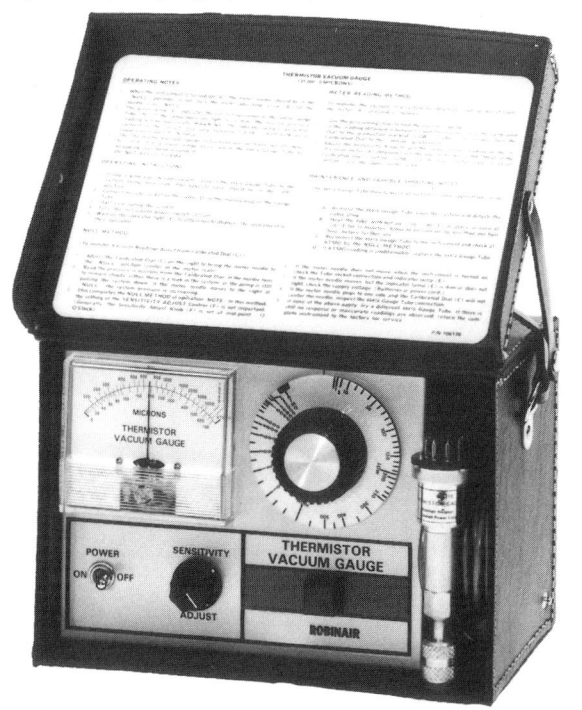

*Figure 11-19. Analog vacuum gauge (Courtesy, Robinair Division, SPX Corporation)*

For oil charging, an oil pump may be used to inject oil into a pressurized system, Figure 11-20. Or, refrigerant may carry the oil from a reservoir when charging, Figure 11-21.

## Charging

Given today's emphasis on energy efficiency, charge levels for some systems are critical. Charging cylinders like the one shown in Figure 11-22 have been commonly used to charge systems.

The graduated shroud, in conjunction with the sight glass, is based on liquid and vapor density calculations. The pressure gauge at the top is coordinated with vertical graduated lines for each refrigerant. Charging cylinders are limited in accuracy (generally ± 4%), partly due to the number of refrigerant scales that will fit on the graduated shroud.

Opening the top valve of the charging cylinder to aid in filling results in venting 3% to 11% of the charge. If the valve is adjusted to fill at a rate of 0.5 lb/min (0.23 kg/min), venting is minimized. Connecting the vent port to an evacuated cylinder or recovery unit eliminates venting entirely.

In automotive systems, the 14-oz can has been used with a can tap valve kit, Figure 11-23. Accuracy and safety are major concerns, so this method of charging is *not* recommended.

Because of its accuracy, the electronic charging scale is popular, Figure 11-24. It measures mass, therefore, it works with all refrigerants.

Electronic scales are typically accurate to ±1 oz for small charges or ±1% for large amounts. Some states have requirements that scales be type certified by a NIST-approved laboratory. It is unfortunate that authorities have singled out the most accurate method of charging refrigeration systems under the guise of protecting the consumer against false weights.

Any method of charging may come under scrutiny if the customer is billed by weight. To avoid this, it may be preferable to charge the customer a flat fee for processing refrigerant within a certain range (e.g., 5 to 10 lb) or buy a type-certified scale.

## Recordkeeping

As with keeping financial records for tax purposes, records of refrigerant purchases and system services are primarily defensive. Refer to the refrigerant leakage provision of EPA Section 608 regulations.

Service Procedures

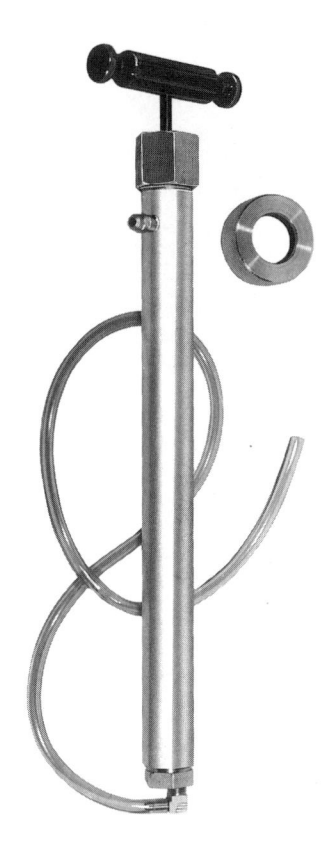

*Figure 11-20. Oil injection pump (Courtesy, Robinair Division, SPX Corporation)*

*Figure 11-21. Oil injector (Courtesy, Robinair Division, SPX Corporation)*

175

The Challenge of Recycling Refrigerants

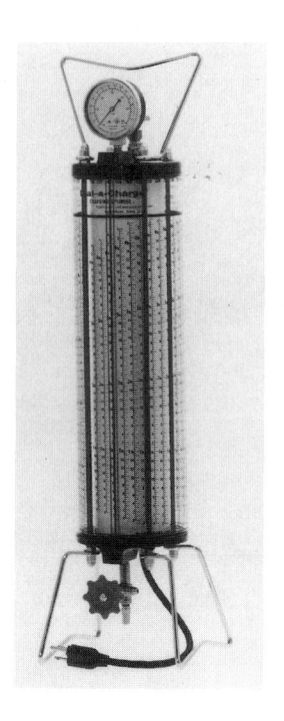

*Figure 11-22. Charging cylinder (Courtesy, Robinair Division, SPX Corporation)*

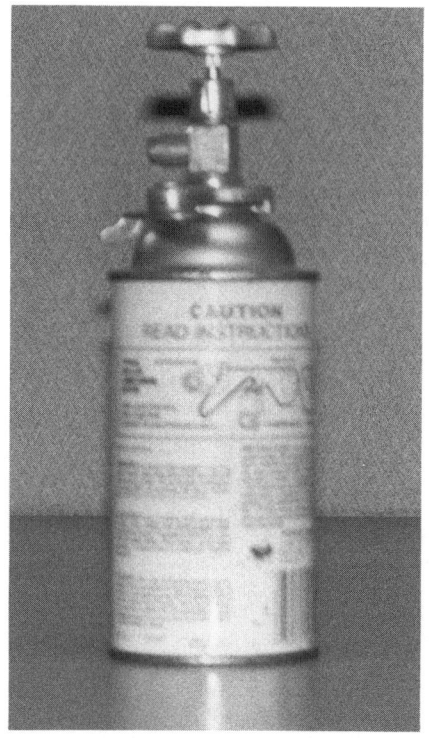

*Figure 11-23. A 14-oz can and tapping valve (Courtesy, TCC-Sercon)*

*Figure 11-24. Electronic charging scale (Courtesy, Robinair Division, SPX Corporation)*

For systems containing more than 50 lb (23 kg) of refrigerant, annual leaks in excess of a predetermined percentage must be repaired or a plan to retrofit or replace the system must be developed within 30 days. The contractor or building owner should keep records for each system in order to prove compliance, if challenged.

In currently drafted California South Coast Air Quality Management District (SCAQMD) Rule 1415 regulations, each system containing more than 50 lb (23 kg) will need to be checked for leak tightness every 30 days by a certified technician. Imagine the documentation required — especially for corrective actions.

Generally, records of refrigerant purchases, how recovered refrigerant is processed, the amount charged into systems (log book), major repairs (particularly for systems containing more than 50 lb or 23 kg), leak test methods, and other necessary documents should be kept for a period of three years. Contact your regional EPA office for the latest recordkeeping recommendations.

## Summary

Despite a huge shift in our views about conserving and recycling refrigerant, the basic elements of service remain much the same. Using low-permeation hoses with shutoff valves and special automotive fittings for R-134a highlight the connecting changes. Refrigerant identification and increased emphasis on leak detection characterize the diagnosing additions.

Recovering and processing refrigerant for reuse, while keeping the refrigerant in the right container to avoid mixing, are major new steps in removing refrigerant. Due to higher service costs, quality work to increase the service interval also gains importance.

High-quality vacuum pumps and electronic micron gauges are used to ensure that systems perform as designed. New electronic scales, which can be used for any refrigerant, help ensure proper charge levels.

While limited records have been kept for very large systems, recordkeeping has not previously been a basic service element. It has become an important discipline for all service work.

# Reference

[1] ASHRAE, *1994 Handbook*, Chapter 6.

## Chapter Twelve

# A Look to the Future

The environmental and human health issues of ozone depletion, global climate change, and indoor air quality place new demands on the entire hvac/r industry, and perhaps even more so, on the users of these mechanical systems. Remember, we *all* depend on refrigeration in one way or another.

Ozone depletion has prompted refrigerant conservation and the search for refrigerants to replace CFCs and HCFCs. Global climate change is eased by the use of more energy-efficient equipment. Engineers, contractors, technicians, and manufacturers face the challenge of retrofitting and replacing CFC-using systems with new refrigerants and lubricants, while maintaining existing systems (some of which may be retrofitted with non-HCFC refrigerants in the future).

Intense competition between system manufacturers is likely to continue. Discriminating system owners will demand improvements in energy and maintenance costs (to offset the capital outlay), as well as improved service.

## Training

Current U.S. EPA Sections 608 (residential-commercial systems) and 609 (motor vehicle systems) rules require technician training and certification. This training provides limited exposure to ozone depletion issues and refrigerant conservation practices; it is by no means adequate to cover all the new demands placed on or abilities required of service technicians.

Service training should be provided on a continuous basis to ensure maximum refrigerant containment; proper cleaning of refrigerants, lubricants, and systems; and knowledge of current technology.

This training may be obtained from system manufacturers, vocational and technical schools, and associations (manufacturer, professional, service). Preferably, voluntary industry programs will identify properly trained technicians so that their skills may be applied if they choose to move to another location.

## Recruiting

The refrigerating industry must attract talented people in order to ensure proper system service in the future. In the past, unions, contractors, and vocational schools have led recruitment efforts. They offered union-sponsored training and apprentice programs, informal contractor training programs, and formal schooling choices. Recently, the refrigerating industry has moved to expand recruiting efforts in order to provide for the future it envisions.

## Tools and Equipment

Service personnel will require more and better tools and equipment in the future. Recycling practices will be applied to current and replacement refrigerants. Leak detectors, refrigerant recovery-recycling equipment, refrigerant-identifying instruments, and indoor air quality instruments will be standard equipment. Increased emphasis will be placed on wider use of existing tools, such as vacuum pumps and charging scales. In summary, technicians must do the things they have always done better, as well as the new things required by environmental concerns and new technology.

## Refrigerants and Lubricants

Just how many refrigerants will be widely used in the future is not known. In the next 15 years, it will not be unusual for individual contractors to deal with more than 10 past, interim, and future refrigerants routinely. Some of these will be blends (zeotropes), and some may be flammable. The task of keeping them all separate will be challenging.

## The Challenge of Recycling Refrigerants: Summary

Chapter 1 discussed the issues surrounding renewable resources, recycling practices, and why people choose or refuse recycling were raised. The treatment of air, land, and stream as jointly held property was used to provide analogy to better understand societal issues.

Chapter 2 covered the historical refrigerant usage and the changes brought about by ozone depletion. The three benign characteristics of non-toxicity, low cost, and long life contributed to careless handling practices that eventually caused the phaseout of chlorofluorocarbon refrigerants.

Chapter 3 detailed the various types, functions, and components of refrigerant recovery units in order to better understand recovery options.

Chapters 4 through 6 concerned used-refrigerant contaminants (oils, acids, particulates, moisture, and air). Options for removing and measuring each contaminant were illustrated.

Chapter 7 described newly raised issues concerning mixed refrigerants, brought about by the option to put more than one type of refrigerant in a system (retrofit) and concerning recycling practices. Efforts to protect the used-refrigerant supply, prevent mixing refrigerants, and analyze for mixed refrigerants, were studied.

Chapter 8 reviewed the options for processing refrigerant for reuse, along with important parameters for choosing between the options. Regulations, equipment manufacturer policies, and industry-recommended practices were important considerations.

Chapter 9 provided assistance for choosing recovery-recycling units for stationary systems. In addition to reviewing ARI Standard 740 test methods and rating parameters, practices were described for optimizing equipment to meet certain application requirements.

Chapter 10 discussed motor vehicle system recycling and retrofitting. Separate service equipment and unique fittings for R-134a have been used to safeguard the used-refrigerant supply.

Chapter 11 explored the basic steps of the service procedure, with emphasis on changes brought about by refrigerant conservation and recycling practices. Diagnosing, removing refrigerant, and recordkeeping steps received special emphasis. Earlier in this chapter, service training, tools, and recruitment were underscored.

## A Final Word

Refrigerants are now viewed as highly valued, reusable products. The challenges of a complete changeout of refrigerants and lubricants, while maintaining the function of these systems during an orderly transition, will demand conservation and recycling. Refrigerant conservation and recycling practices will be used with all current, interim, and future refrigerants.

Too often, engineers have responded to technological challenges by being over-prescriptive. In this book, the emphasis has been placed on raising issues, discussing options, and presenting relevant information that helps paint the "big picture" of

## The Challenge of Recycling Refrigerants

what needs to be done. Armed with this understanding, only you can use the tools, equipment, and practices to protect the environment.

Technicians have experience engineers do not possess. Industry members' hats are off to you, the foot soldiers, as you go out to do what society demands for a safe environment. Nonetheless, spot shortages of refrigerants are likely to occur during this transition. In addition, issues of energy efficiency and indoor air quality will pose further demands on the air conditioning and refrigerating industry.

This industry has and will continue to respond to those challenges.

# Index

## A

Acid 45
Air 73

## B

Boiling Point 96

## C

Chlorofluorocarbons 9, 11
   tax 14
Clean Air Act 12
   Section 608 14, 119
   Section 609 14, 106, 149
Compressors 20
  automotive recovery unit 21
  burnout 107
  oil reservoir 49
  oil-less 22, 50
Condenser 37
Consumable Products 3

## E

Environment 2
Evacuation 73, 109, 169
   service connections 75
   vacuum pump. *See* Vacuum Pump

## F

Filter-Driers 55
   change indicators 59, 63
   changing 70
   in recycling equipment 58
   liquid 59
   loose-fill 56, 59
   suction 59
Flow Controls 30
   capillary tube 30, 31
   low-side float 33
   thermostatic expansion valve (TXV) 31

## G

Global Warming 14, 15

## H

Heat Exchangers 17, 18
   coaxial coil 20
Hourmeter 63

## I

In-Situ Mass Flowmeter 65

# M

Metering Devices. *See* Flow Controls
Moisture 55
   calculations 68
   indicators 61
   laboratory sampling 71
   removal 55
      desiccants 56
      filter-driers. *See* Filter-Driers
   removing free water 70
Montreal Protocol 12
Motor Vehicle Air Conditioners 137
   flushing 146
   flushing survey 140
   mixed refrigerants 149
   recycling equipment 142, 144
   refrigerant identification 147
   retrofitting vehicles 138

# N

Non-Condensables 76
   analysis 86
   in a storage container 82
   measurement 80
   oxygen sensor 77
   partial pressure 76, 80
   purging 77, 82
      automatic purge valve 77
      desiccant adsorption method 86
      efficiency 84
      manual air-purge valve 78

# O

Oil 45
   analysis 53
   disposing of 52
   draining 49
   removal 46
   separation 46
Oil Separator 21
   canister 46
   commercial 49

   heat exchanger 46
   liquid-phase 50
Ozone Depletion 10, 11
   timeline 11

# R

Recover 17
Recovery Equipment 17
   automotive air conditioner 30, 38
   certification 35
   clearing 35, 37
   contaminant levels 117
   operation controls 33
   portability 117
   rating parameters 115
   recovery time 118
   requirements 116
   vacuum 34
   weight scale 70
Recruiting 180
Recycling 1, 2
   mindset 1, 2, 5
Recycling Equipment 58
   cleaning capabilities 118
   contaminant levels 117
   filter replacement 118
   filter-driers. *See* Filter-Driers: in recycling equipment
Refrigerant 16
   analysis 100
   blend 89
   future 180
   life cycle 16
   liquid properties 99
   mixed 89
      preventing 92
   processing 105
   reclaim 112
   recovered 107, 109
   recycled 108, 111
   used 24, 46, 90, 109
      motor vehicle air conditioners 93
      selling 92, 105
   vapor sample 97
Renewable Products 4

# Index

## S

Service Procedures 161
   charging 174
   diagnosing 162
   recordkeeping 174
   removing refrigerant 167
   repairing 169
Storage Containers 24
   acidic 52
   ARI Guideline K 27
   cleaning procedure 51
   cooling 39
   filling 27, 28
   labeling 27
   retesting 28

## T

Training 179

## U

Used Materials 6
   compressor oil 10

## V

Vacuum Pump 22, 40, 73
Venting 10
Volumetric Flowmeter 63

## Other Titles Offered by BNP

Fine Tuning Air Conditioning Systems and Heat Pumps
Refrigeration Reference Notebook
A/C and Heating Reference Notebook
Practical Cleanroom Design, Revised Edition
Managing People in the Hvac/r Industry
Troubleshooting and Servicing Modern Air Conditioning and Refrigeration Systems
Fire Protection Systems
Piping Practice for Refrigeration Applications
Born to Build: A Parent's Guide to Academic Alternatives
Refrigeration Fundamentals
Technician's Guide to Certification
Plumbing Technology
Power Technology
Refrigeration Licenses Unlimited, Second Edition
A/C Cutter's Ready Reference
How to Solve Your Refrigeration and A/C Service Problems
Blueprint Reading Made Easy
Starting in Heating and Air Conditioning Service
Legionnaires' Disease: Prevention & Control
Schematic Wiring, A Step-By-Step Guide
Sales Engineering
Hydronics
How to Design Heating-Cooling Comfort Systems
Industrial Refrigeration
Modern Soldering & Brazing Techniques
Inventing from Start to Finish
Indoor Air Quality in the Building Environment
Electronic HVAC Controls Simplified
Heat Pump Systems and Service
Ice Machine Service
Troubleshooting and Servicing A/C Equipment
Stationary Engineering
Medium and High Efficiency Gas Furnaces
Water Treatment Specification Manual, Second Edition
The Four R's: Recovery, Recycling, Reclaiming, Regulation
SI Units for the HVAC/R Professional

TO RECEIVE A FREE CATALOG, CALL

## 1-800-837-1037

**BNP**
**BUSINESS NEWS**
**PUBLISHING COMPANY**
Troy, Michigan
USA

# Troubleshooting at its best!

Your work environment is a complex one. Not only must you stay abreast of all new advances in technology, improved service techniques, and environmental regulations, you must have a sound understanding of how many different systems work.

*Troubleshooting & Servicing Modern Air Conditioning & Refrigeration Systems*, a new book from BNP Technical Books, was written to give service technicians all the information needed to accurately diagnose and solve various system problems. This book will give you a greater understanding of refrigeration and air conditioning while exploring more complex topics and detailed troubleshooting procedures. It is a valuable tool that will give you a helping hand at the jobsite or in the shop.

Inside you'll find the important topics of today — the changes affecting the refrigeration and air conditioning industries with an emphasis on the phase-out of CFC and HCFC refrigerants. You'll get detailed information on the most current leak detection methods, venting regulations, alternative refrigerants, and retrofit guidelines. In addition, you'll get important refrigerant changeover guidelines for the following conversions: R-12 to R-134a; R-12 to MP39; and R-502 to SUVA HP80.

Hardcover • 282 pages

## Troubleshooting & Servicing Modern Air Conditioning & Refrigeration Systems
## Order Today! 1-800-837-1037

**BNP**
Technical Books
The books of choice.

# The ultimate hvac/r reference set!
# Reference Notebook Set

**Air Conditioning Heating & Refrigeration Dictionary**
Second Edition
Tim Zurick

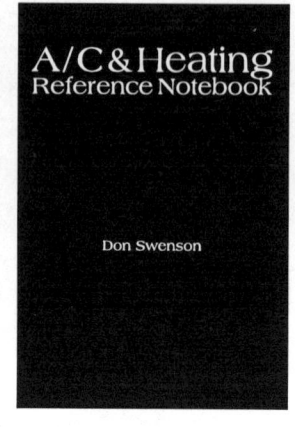

**A/C & Heating Reference Notebook**
Don Swenson

**Refrigeration Reference Notebook**
Don Swenson

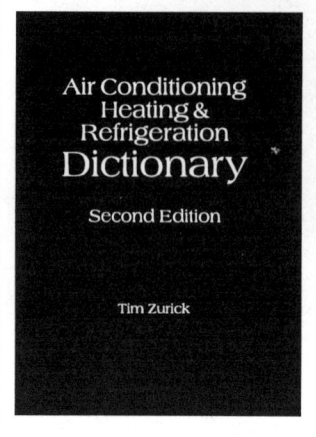

It's new! It's improved! The all-time BNP classic *Reference Notebook Set* has been updated to meet the needs of today's service technician.

With the *Reference Notebook Set*, you can quickly locate important definitions, capillary tube length conversion charts, cooling load factors, pipe sizing tables, and much more.

What makes this set so good is that each book can fit into your pocket, giving you instant information that makes your job easier and more efficient. The set consists of:

• **Air Conditioning, Heating, and Refrigeration Dictionary, Second Edition** - An updated, alphabetical listing of all relevant hvac/r terms and their definitions.

• **A/C & Heating Reference Notebook** - Brand new and up-to-date numerical information on heating and cooling loads, humidification, ducts and blowers, piping and tubing, as well as measurement conversions, all found in table form for quick reference.

• **Refrigeration Reference Notebook** - Completely revamped with updated information on heat load factors, refrigerant properties, pipe and tubing sizing, wire and metal sizing, blower data, conversions, and more!

Whether you are new to the hvac/r fields or have years of experience, the *Reference Notebook Set* is a must have.

Each book can be ordered separately. Call for more information.

## Order Today! 1-800-837-1037

ASHEVILLE-BUNCOMBE
TECHNICAL COMMUNITY COLLEGE

3 3312 00046 9981

TP 492.7 .M343 1995

Manz, Kenneth W., 1951-

The challenge of recycling
 refrigerants

DISCARDED

JUN 2 6 2025